THE COMMONWEALTH AND INTERNATIONAL LIBRARY

Joint Chairmen of the Honorary Editorial Advisory Board

SIR ROBERT ROBINSON, O.M., F.R.S., LONDON

DEAN ATHELSTAN SPILHAUS, MINNESOTA

Publisher: ROBERT MAXWELL, M.C., M.P.

SELECTED READINGS IN PHYSICS

General Editor: D. TER HAAR

KINETIC THEORY

Volume 2: Irreversible Processes

KINETIC THEORY

Volume 2   IRREVERSIBLE PROCESSES

by

S. G. BRUSH

PERGAMON PRESS

OXFORD · LONDON · EDINBURGH · NEW YORK
TORONTO · SYDNEY · PARIS · BRAUNSCHWEIG

Pergamon Press Ltd., Headington Hill Hall, Oxford
4 & 5 Fitzroy Square, London W.1

Pergamon Press (Scotland) Ltd., 2 & 3 Teviot Place, Edinburgh 1

Pergamon Press Inc., 44–01 21st Street, Long Island City, New York 11101

Pergamon of Canada, Ltd., 6 Adelaide Street East, Toronto, Ontario

Pergamon Press (Aust.) Pty. Ltd., 20–22 Margaret Street, Sydney,
    New South Wales

Pergamon Press S.A.R.L., 24 rue des Écoles, Paris 5ᵉ

Vieweg & Sohn GmbH, Burgplatz 1, Braunschweig

Printed in Great Britain by Bell and Bain Ltd., Glasgow

To Denise

Contents Volume 2

Volume 1

A*

Preface

THIS is the second volume of a collection of fundamental papers on the kinetic theory of gases; it includes the two papers by Maxwell and Boltzmann in which the basic equations for transport processes in gases are formulated, together with the first derivation of Boltzmann's " H-theorem " and a sample of the later discussion of this theorem and the problem of irreversibility.

Whereas in Volume 1 of this series we restricted ourselves to reprinting short papers of a relatively elementary character, or gave only extracts from longer works, the present volume is almost entirely occupied by two rather long and detailed memoirs, one of which is published here in English for the first time. It is hoped that serious students of physics (at the advanced undergraduate or beginning postgraduate level) will benefit by a close study of these works. Although some of the results of Maxwell's and Boltzmann's investigations have been incorporated into modern textbooks, the derivations are usually presented in a very condensed form. Boltzmann's exposition goes to the opposite extreme and is perhaps unduly elaborate and repetitious; nevertheless one can acquire a deeper understanding of the subject if he is willing to spend the time to follow through the details of the derivations (as is indicated by the continued and deserved popularity of Tolman's monumental book, *The Principles of Statistical Mechanics*. It may be noted that Maxwell's theory, which for many years was considered somewhat irrelevant to physics because of its reliance on an artificial molecular model, is now recognized to be of great value in the mathematical treatment of transport phenomena; the Maxwellian inverse fifth-power force law provides an idealized limiting case that plays the same role in transport theory as the ideal gas does in the study of equilibrium thermodynamic properties.

For readers who wish to acquire a bird's-eye view of the subject without making an intensive study of the technical parts, we recommend the following: Introduction; Selection 1, first 7 pages and last 10 pages; Selection 2, first 5 pages; sections immediately following equations (16) and (25), and the discussion of entropy in the last 3 pages; Selection 3, first 5 pages; all of Selection 6; all of Selection 10.

Addendum to Volume 1

After the first volume of this series was prepared, historical research was published which showed that the relation between pressure and volume of a gas, commonly known as Boyle's Law, should be credited to Henry Power as well as to Towneley and Boyle. See C. Webster, *Nature* **197,** 226 (1963), *Arch. Hist. Exact Sci.* **2,** 441 (1965), and I. B. Cohen, *Nature* **204,** 618 (1964).

Part 1

Introduction

THE works reprinted in the first volume of this series were concerned with establishing the fundamental nature of heat and of gases. By about 1865 this question seemed to have been definitely settled by the general adoption of thermodynamics and the kinetic theory of gases. Scientists could now turn their attention to working out the detailed consequences of these theories and comparing them with experiments.

Knowing the empirical laws relating pressure, volume, and temperature, many scientists had discovered for themselves the kinetic explanation based on the collisions of molecules with the walls of the container. The ideal gas laws could be deduced by assuming that the space occupied by the molecules themselves is negligible compared to that total volume of the gas; it was only a natural extension of the theory to try to explain deviations from the ideal gas laws by dropping this assumption, although it was not until 1873 that J. D. van der Waals found the most fruitful mathematical formulation of this extension. But accurate experimental data on viscosity, heat conductivity, and diffusion were not available until well after 1850, and the first kinetic theorists had to make their calculations of these properties without the advantage of knowing the answer beforehand. Moreover, the type of assumption that one makes about short-range intermolecular forces determines the first approximation to the transport coefficients, whereas it comes in only in the second approximation to the equilibrium properties. Of course, such circumstances made it all the more convincing when experiments confirmed the theoretical predictions; for example, Maxwell's discovery that the viscosity of a gas is independent of density (Selection 10, Vol. 1) was very important in establishing the kinetic theory.

3

One may distinguish two types of irreversible process: in the first, there is a continuous flow of mass, momentum, or energy resulting from variations in concentration, velocity, and temperature imposed externally; in the second, a gas that is initially in a non-equilibrium state spontaneously moves toward equilibrium in the absence of external interference. (The two types are, of course, intimately related, as shown by Maxwell's theory of relaxation processes (Selection 1) and by modern work on the " fluctuation-dissipation theorem "—transport coefficients can be expressed in terms of the relaxation of fluctuations, and conversely.)

Following the introduction of the " mean-free-path " concept by Clausius in 1858 (Vol. 1, Selection 9) and Maxwell's theory of transport processes based on it (Vol. 1, Selection 10) there was a large amount of theoretical work by such scientists as O. E. Meyer, J. Stefan, P. G. Tait, G. Jäger, and J. H. Jeans.† This work was devoted to refining the mean-free-path theory and applying it to various phenomena. Maxwell himself realized almost immediately that the mean-free-path method was inadequate as a foundation for kinetic theory, even though it might be useful for rough approximations, and while he revised his own original treatment of diffusion and heat conduction he did not publish any further works based on this method.‡ Instead he developed the much more accurate technique based on transfer equations, described in Selection 1.

Maxwell's transfer equations provide a method for computing the rate of transport of any quantity such as mass, momentum, and energy which can be defined in terms of molecular properties, consistently with the macroscopic equations such as the Navier–Stokes equation of hydrodynamics. However, the actual calculation of transport coefficients depends in general on a knowledge of

---

† See the works by Boltzmann (1896–8), Brush (1962), Jeans (1925) Kennard (1938) and Loeb (1934), cited in the Bibliography.

‡ For diffusion theory, see Furry (1948). It is hoped that Maxwell's manuscripts on heat conduction and other topics in kinetic theory will be published shortly.

the velocity–distribution function, which is no longer assumed to be the same as the equilibrium Maxwell distribution; yet there is no direct way to compute this function. It was not until 1916 that Sydney Chapman worked out a systematic method for calculating the velocity–distribution function and all the transport coefficients of gases by means of the transfer equations. However, Maxwell discovered that in one special case, when the force between the molecules varies inversely as the fifth power of the distance, the distribution function enters the equations in such a way that one can calculate the transport coefficients without knowing it. In this case it turns out that the viscosity coefficient is directly proportional to the temperature, whereas in the case of hard spheres it is proportional to the square root of the temperature. Maxwell's own measurements of the viscosity of air, which he had been carrying out at about the same time, led him to believe that the former result is in agreement with experiment.† The inverse fifth-power force, which is an exactly soluble case for purely mathematical reasons, therefore seemed to be at the same time a good model for real gases, and Maxwell adopted it provisionally as the basis for his later work.‡

Boltzmann

Ludwig Boltzmann (1844–1906), whose writings occupy the major part of this volume, was born and educated in Vienna, and taught there most of his life. He is best known for his work on the kinetic theory of gases and statistical mechanics. His equation for the velocity–distribution function of molecules (first derived in the work reprinted here as Selection 2) has been used extensively to study transport properties such as viscosity, thermal conductivity, and diffusion. Boltzmann's equation is actually equivalent to

---

† J. C. Maxwell, *Phil. Trans. Roy. Soc. London* **156**, 249 (1866).

‡ J. C. Maxwell, *Trans. Cambridge Phil. Soc.* **12**, 547 (1879); see also Boltzmann (1896), Chapter III. The later work of Chapman and Enskog showed that thermal diffusion, which takes place in systems of molecules with any general force law, happens to be absent in the special case of inverse fourth-power forces.

Maxwell's set of transfer equations, and its solution presents exactly the same difficulties. Just as Chapman solved Maxwell's equations, so did David Enskog (1917) solve Boltzmann's equation in the general case, and the Maxwell–Boltzmann–Chapman–Enskog theory constitutes the completion of the classical kinetic theory of low-density gases. The basic theory has remained essentially unchanged since 1917; with appropriate modifications, it forms the basis for many modern theories of liquids, solids, plasmas, and neutron transport.

Though Boltzmann himself did not obtain the complete solution to his equation for the distribution function, he did deduce from it one important consequence, which later[†] came to be known as the H-theorem (see the last part of the first section of Selection 2). The H-theorem attempts to explain the irreversibility of natural processes by showing how molecular collisions tend to increase entropy; any initial distribution of molecular positions and velocities will almost certainly evolve into an equilibrium state, in which the velocities are distributed according to Maxwell's law.

The Maxwell velocity distribution law was also generalized by Boltzmann so as to apply to physical systems in which interatomic forces and external fields must be taken into account.[‡] The so-called Boltzmann factor, $e^{-E/kT}$, which gives the relative probability of a configuration of energy E at a temperature T, provides the basis for all calculations of the equilibrium properties of matter from the molecular viewpoint. Boltzmann further showed that the entropy of a system in any physical state may be calculated from its probability by counting the number of molecular configurations corresponding to that state.[§] This result, and the

---

[†] Boltzmann himself originally used the letter E, and did not change to H until 1895; the first use of H for this quantity was apparently by S. H. Burbury, *Phil. Mag.* **30**, 301 (1890). A rumour has circulated among modern physicists that H was intended to be a capital eta, the letter eta having been used for entropy by Gibbs and other writers [see S. Chapman, *Nature* **139**, 931 (1937)]. This can hardly be true if in fact the usage was originated by Burbury, since he makes no such suggestion in his paper.

[‡] L. Boltzmann, *Wien. Ber.* **58**, 517 (1868).

[§] L. Boltzmann, *Wien. Ber.* **76**, 373 (1877).

Boltzmann–Stefan formula for the energy of black-body radiation,[†] aided Max Planck in his development of quantum theory in 1900.[‡]

Boltzmann's work was followed with considerable interest by British scientists, and conversely Boltzmann helped to convert German scientists to Maxwell's electromagnetic theory.[§] However, he did not always find support for his own theories in Germany. In the 1890's he had to defend the kinetic theory, and even the existence of atoms, against the attacks of Wilhelm Ostwald, Pierre Duhem, Ernst Mach and others. Boltzmann thought himself to be on the losing side in this battle, and feared that " Energetics " and the " reaction against materialism " would obliterate his life-work.[‖] However, within a few years of his suicide in 1906, the existence of atoms had been definitely established by experiments such as those on Brownian motion, and even Ostwald had to admit that he had been wrong.[¶]

Dissipation of Energy

The modern idea of irreversibility in physical processes is based on the second law of thermodynamics in its generalized form. As is well known, the second law was first enunciated by Sadi Carnot in 1824 in his memoir on the efficiency of steam engines,[††] and the

---

[†] Stated as an empirical generalization by J. Stefan, *Wien. Ber.* **79,** 391 (1879), and derived from thermodynamics and Maxwell's electromagnetic theory by L. Boltzmann, *Ann. Phys.* **22,** 291 (1884).

[‡] See M. J. Klein, *Arch. Hist. Exact Sci.* **1,** 459 (1962); *The Natural Philosopher* **1,** 83 (1963).

[§] L. Boltzmann, *Vorlesungen über Maxwells Theorie der Elektricität und des Lichts,* J. A. Barth, Leipzig, 1891 and 1893.

[‖] See Boltzmann (1896–8), pp. 13–17, 23–8, 215–16 of the English translation.

[¶] Compare W. Ostwald, *Nature* **70,** 15 (1904) (before) and *Grundriss der allgemeinen Chemie,* Engelmann, Leipzig, 4th ed., 1909, Vorbericht (after). [English translation of the latter published as *Outlines of General Chemistry,* Macmillan, London, 1912.]

[††] See Carnot (1824), cited in Bibliography.

problem of obtaining the maximum amount of work from a given amount of fuel seems to have provided the motivation for many of the nineteenth-century researches on heat and its transformations.

Although one can find scattered statements in the technical literature before 1850 to the effect that something is always lost or dissipated when heat is used to produce mechanical work, it was not until 1852 that William Thomson (later Lord Kelvin) asserted the existence of "A universal tendency in nature to the dissipation of mechanical energy ".† At the same time he mentioned his conclusion that " within a finite period of time past, the earth must have been, and within a finite period of time to come the earth must again be, unfit for the habitation of man as at present constituted, unless operations going on at present in the material world are subject ".

The consequences of Thomson's Dissipation Principle were elaborated further by Hermann von Helmholtz in a lecture 2 years later, in which he described the final state of the universe: all energy will eventually be transformed into heat at uniform temperature, and all natural processes must cease; "the universe from that time forward would be condemned to a state of eternal rest ". Thus was made explicit the concept of the "heat death" of the universe.‡

The modern statement of the Dissipation Principle involves the notion of " entropy ", introduced by Clausius in 1865.§ It should be emphasized that no new physical content was being added to the second law in this formulation of Clausius, yet the mere act of giving a new short name to a physical quantity that had previously been represented only by mathematical formulae and awkward circumlocutions had an undeniable influence on the subsequent history of the subject. Entropy is taken from the Greek $\tau\rho o\pi\dot{\eta}$, meaning transformation, and was intentionally chosen to have a resemblance to the word energy. The two laws of thermodynamics are then stated as follows:

---

† See Kelvin (1852).

‡ H. von Helmholtz, *Popular Scientific Lectures* (translated from German and edited by M. Kline), Dover Publications, New York, 1962, p. 59.

§ R. Clausius, *Ann. Phys.* **125**, 353 (1865).

1. The energy of the universe is constant.
2. The entropy of the universe tends toward a maximum.

Before returning to the role of energy dissipation in the history of the kinetic theory, we should mention the primary use which Thomson made of his Dissipation Principle. One of the major goals of mathematical physics in the eighteenth century had been to deduce the stability of the solar system from Newtonian mechanics, and by the beginning of the nineteenth century this goal had apparently been reached by the work of such scientists as Lagrange, Laplace and Poisson. Consequently it was generally believed that the earth had remained at the same average distance from the sun for an indefinitely long time in the past, and that physical conditions on the surface of the earth had been roughly the same as they are now for countless millions of years. This assumption was the basis for the " Uniformitarian " theory in geology, advocated by Hutton and Lyell, which gradually replaced the earlier " Catastrophist " doctrine. According to the Uniformitarians, the present appearance of the earth's surface was to be explained as a result of physical causes, like erosion, whose operation can still be observed, rather than by postulating catastrophic upheavals in the past such as the Flood.[†] Likewise, the Darwin–Wallace theory of biological evolution by natural selection assumes that the origin and development of species could have taken place by gradual changes over a very long period of time, during which physical conditions did not change very much, and indeed Darwin relied heavily on Uniformitarian geology in establishing his theory.[‡] It was just this assumption of constant physical conditions that Thomson attacked, using the Dissipation Principle in conjunction with Fourier's theory of heat conduction. He showed that if one assumed that the earth had once been

[†] C. C. Gillispie, *Genesis and Geology*, Harvard University Press, 1951; A Wolf, *A History of Science, Technology, and Philosophy in the* 18*th Century* Macmillan, New York, 2nd edition 1952, Vol. I, Chap. XV; S. J. Gould, *Amer. J. Sci.* **263**, 223 (1965).

[‡] See " On the Imperfection of the Geological Record " which is Chapter IX in the first edition of the *Origin of Species* (1859) and Chapter X in later editions.

very hot and was slowly losing its heat by conduction outward from the centre, and that no other sources of heat are present, then it follows from experimental data on the present rate of heat loss and certain plausible assumptions about the conductivity of the interior that present conditions can have lasted no more than about 20 million years. Before that time the temperature of the entire earth must have been so high that the whole globe was liquid. Therefore Uniformitarian geology, whose advocates had assumed constant conditions over periods of hundreds of millions of years, must be wrong.†

There followed a bitter controversy between physicists (led by Thomson and P. G. Tait) and geologists (led by T. H. Huxley) about the validity of various methods for estimating the age of the earth.‡ Religious views played some role in this dispute, for Thomson believed that he had discovered a mathematical proof that there must have been a creation by solving Fourier's heat equation.§ If he could discredit Uniformitarian geology, he would also have removed one of the major supporting arguments from the theory of evolution by natural selection, a theory which he considered " did not sufficiently take into account a continually guiding and controlling intelligence ".‖ Darwin himself was well aware of the fact that Thomson's attack weakened his own theory, and one historian of evolution has noted that the physicists " forced Darwin, before his death, into an awkward retreat which mars in some degree the final edition of the *Origin* ".¶

---

† W. Thomson, *Trans. Roy. Soc. Edinburgh* **13,** 157 (1864); *Proc. Roy. Soc. Edinburgh* **5,** 512 (1866); *Trans. Glasgow. Geol. Soc.* **3,** 1, 215 (1871); *A Source Book in Astronomy* (ed. H. Shapley and H. E. Howarth), McGraw-Hill, New York, 1929.

‡ T. H. Huxley, *Q. J. Geol. Soc. London,* **25,** xxviii (1869); anonymous article in the *North British Review,* **50,** 406 (1869), attributed to P. G. Tait by Darwin, in a letter to J. D. Hooker, in *More Letters of Charles Darwin,* Appleton, New York, 1903, **1,** 313–14.

§ S. P. Thompson, *The Life of William Thomson, Baron Kelvin of Largs,* Macmillan, London, 1910, **1,** 111.

‖ W. Thomson, *British Assn. Rept.* (1871) p. lxxxiv; *Nature* **4,** 262 (1871).

¶ L. Eiseley, *Darwin's Century,* Doubleday, Garden City, 1958, p. 245.

The dispute was finally settled in favour of the geologists by the discovery of radioactivity, which not only provided a source of heat which Thomson had not taken into account in his calculations, but also made it possible to determine the age of rocks fairly accurately. Thomson himself lived long enough to see the publication of these studies which showed that the earth must be at least 2 thousand million years old.†

Cultural Influence of the Dissipation of Energy

The English philosopher Herbert Spencer was one of the first writers to take up the idea of dissipation of energy and incorporate it into a general system of philosophy.‡ Despite the apparently pessimistic consequences of the heat death, Spencer manages to draw from the second law " a warrant for the belief, that Evolution can end only in the establishment of the greatest perfection and the most complete happiness ". He also predicts that after the motion of all the stars has become equilibrated and degraded to heat energy, there will be a process of concentration under the action of gravity, followed by dispersion, and an infinite sequence of alternate eras of Evolution and Dissolution.

The most remarkable application of the Dissipation Principle was made by the American historian Henry Adams. In three essays, later published under the title *The Degradation of the Democratic Dogma*, he argued that a science of history could be based on the general properties of energy as discovered by the physical sciences.§ He saw a general process of degradation and deterioration in human history, and noted that the cheerful

---

† R. J. Strutt, *Proc. Roy. Soc.* London **A76**, 88 (1905); B. B. Boltwood, *Amer. J. Sci.* **23**, 77 (1907); see also E. N. da C. Andrade, *Rutherford and the Nature of the Atom*, Doubleday, Garden City, 1964, p. 80.

‡ H. Spencer, *First Principles*, London, 1870, pp. 514–17.

§ H. Adams, " The Tendency of History " (1894); " A letter to American Teachers of History " (1910); " The Rule of Phase applied to History " (1909), reprinted with an Introduction by Brooks Adams in *The Degradation of the Democratic Dogma*, Macmillan, New York, 1919.

optimism inspired by Darwin had given way to *fin-de-siècle* pessimism toward the end of the nineteenth century. To the historian, the concept of entropy " meant only that the ash-heap was constantly increasing in size ".

Aside from these and a few other examples, the direct influence of the Dissipation Principle on European thought in the nineteenth century was remarkably small. It is only at the end of the century, and in the first part of the twentieth century, that one finds an increasing number of references to the second law of thermodynamics, and attempts to connect it with general historical tendencies. The writings of Sir James Jeans and Sir Arthur Eddington in the 1920's and 1930's have now made the " heat death " an integral part of the modern educated layman's knowledge of cosmology.†

The Statistical Nature of the Second Law

Naturally not everyone was satisfied with the pessimistic consequences that William Thomson and others believed to follow from the second law, and there were many attempts to disprove or circumvent them. The best-known scheme for this purpose is based on the concept of " Maxwell's Demon ", originally invented in order to illustrate the statistical nature of irreversibility. The Demon was conceived in discussions among Maxwell, Tait and Thomson; he was first described in a letter from Maxwell to Tait in 1867, and publicly introduced in Maxwell's *Theory of Heat* published in 1871.‡ He is " a being whose faculties are so sharpened that he can follow every molecule in its course ". A vessel filled with gas is divided into two parts by a partition with a small hole, and the demon opens and closes the hole so as to allow

---

† A. S. Eddington, *The Nature of the Physical World*, Cambridge University Press, 1928; J. H. Jeans, *The Universe Around Us*, Cambridge University Press, 1933.

‡ See C. G. Knott, *Life and Scientific Work of Peter Guthrie Tait*, Cambridge University Press, 1911, pp. 213–14; J. C. Maxwell, *Theory of Heat*, Longmans, London, 1871, Chapter XXII.

only the swifter molecules to pass through in one direction, and only the slower ones in the other. He thus produces a temperature difference without any expenditure of work, violating the second law. The point of this imaginary construction is that the second law does not necessarily apply to events on a molecular level, if a being with sufficient knowledge about the details of molecular configurations is present to manipulate things.

Boltzmann's *H*-theorem shows the statistical nature of the second law in a more quantitative though less picturesque manner. In his proof of the theorem (Selection 2), Boltzmann first introduced a definition of entropy in terms of the molecular velocity distribution, then showed that if certain assumptions are granted, the entropy as thus defined will always increase as a result of collisions among molecules. The assumptions are of two types: first, statistical assumptions about the random distribution of the velocities of two molecules *before* they collide; and second, mechanical assumptions about the existence of " inverse " collisions. The first assumption is needed in order the calculate the number of collisions of various types; the second is needed in order to balance the effects of some collisions against others in which the initial and final configurations are interchanged, in order to obtain a final result having the desired mathematical properties. The limitations on the validity of these assumptions were not explicitly recognized by Boltzmann until later, and in fact the *H*-theorem inspired a large amount of controversy during the late nineteenth century and afterwards.† Roughly speaking, what the *H*-theorem proves is not so much a property of real physical systems as a property of our information about those systems. The equilibrium Maxwell distribution is the most random one possible, in the sense that it represents the minimum amount of information. If we start with partial information about molecular velocities (so that the distribution is not Maxwellian) then this information will be gradually lost as collisions occur. On the other hand, if we had started with complete information and

† See P. and T. Ehrenfest (1911); ter Haar (1954); Dugas (1959); Bryan (1891–4).

could follow the motions as determined by the laws of (classical) mechanics, no information need ever be lost in principle, and the entropy would remain constant.†

Thomson's 1874 paper on the dissipation of energy (Selection 3) contains a substantial part of the modern interpretation of irreversibility, though it is seldom cited. Thomson notes the contrast between " abstract dynamics ", which is reversible, and " physical dynamics ", which is not, and shows how this " reversibility paradox " may be explained by taking account of the large number of molecules involved in physical processes.

The reversibility paradox is usually attributed to Josef Loschmidt, who mentioned it very briefly in the first of four articles on the thermal equilibrium of a system of bodies subject to gravitational forces.‡ Loschmidt was attempting to show in these papers that equilibrium was possible without equality of temperature; in this way he hoped to demonstrate that the heat death of the universe is not inevitable. He claimed that the second law could be correctly formulated as a mechanical principle without reference to the sequence of events in time; he thought that he could thus " destroy the terroristic nimbus of the second law, which has made it appear to be an annihilating principle for all living beings of the universe; and at the same time open up the comforting prospect that mankind is not dependent on mineral coal or the sun for transforming heat into work, but rather may have available forever an inexhaustible supply of transformable heat ". After attacking Maxwell's conclusion that the temperature of the gas in a column should be independent of height, and proposing a model which supposedly violates this law, he says that in any system " the entire course of events will be retraced if at some instant the velocities of all its parts are reversed ". Loschmidt's application of this reversibility principle to the validity of the second law is somewhat obscurely stated, but Boltzmann quickly got the point and immediately published a reply (Selection 4) in which he gave a thorough discussion of the reversibility

† See Tolman (1938) Chapter VI.
‡ J. Loschmidt, *Wien. Ber.* **73**, 128, 366 (1876); **75**, 287; **76**, 209 (1877).

paradox, ending up with a conclusion very similar to that of Thomson.

The Eternal Return and the Recurrence Paradox

The notion that history repeats itself—that there is no progress or decay in the long run, but only a cycle of development that always returns to its starting point—has been inherited from ancient philosophy and primitive religion. It has been noted by some scholars that belief in recurrence, as opposed to unending progress, is intimately connected with man's view of his place in the universe, as well as with his concept of history. Starting, in most cases, from a pessimistic view of the present and immediate future, it denies the reality or validity of human actions and historical events by themselves; actions and events are real only insofar as they can be understood as the working out of timeless archetypal patterns of behaviour in the mythology of the society. This attitude is said to be illustrated in classical Greek and Roman art and literature, where there is no consciousness of past or future, but only of eternal principles and values. By contrast the modern Western view, as a result of the influence of Christianity, is deeply conscious of history as progress toward a goal.† Nevertheless, the cyclical view has by no means died out, and can easily be recognized in the persistent tendency to draw historical analogies and comparisons.

The suggestion that eternal recurrence might be proved as a theorem of physics, rather than as a religious or philosophical doctrine, seems to have occurred at about the same time to the German philosopher Friedrich Nietzsche and the French mathematician Henri Poincaré. Nietzsche encountered the idea of recurrence in his studies of classical philology, and again in a book

---

† M. Eliade, *The Myth of the Eternal Return*, Pantheon Books, New York, 1954; A. Rey, *Le Retour Eternel et la Philosophie de la Physique*, Flammarion, Paris, 1927; J. Baillie, *The Belief in Progress*, Oxford University Press, 1950, §10; P. Sorokin, *Social and Cultural Dynamics*, American Book Co., New York, 1937, Vol. II, Chap. 10.

by Heine.[†] It was not until 1881 that he began to take it seriously, however, and then he devoted several years to studying physics in order to find a scientific foundation for it.[‡] Poincaré, on the other hand, was led to the subject by his attempts to complete Poisson's proof of the stability of the solar system. Both Nietzsche and Poincaré were trying, though in very different ways, to attack the " materialist " or " mechanist " view of the universe.

Nietzsche's " proof " of the necessity of eternal recurrence (written during the period 1884–88 but not published until after his death in 1900) is as follows: " If the universe has a goal, that goal would have been reached by now " since the universe, he thinks, has always existed; the concept of a world " created " at some finite time in the past is considered a meaningless relic of the superstitious ages. He absolutely rejects the idea of a " final state " of the universe, and further remarks that " if, for instance, materialism cannot consistently escape the conclusion of a finite state, which William Thomson has traced out for it, then materialism is thereby refuted ". He continues:

> If the universe may be conceived as a definite quantity of energy, as a definite number of centres of energy—and every other concept remains indefinite and therefore useless—it follows therefrom that the universe must go through a calculable number of combinations in the great game of chance which constitutes its existence. In infinity, at some moment or other, every possible combination must once have been realized; not only this, but it must have been realized an infinite number of times. And inasmuch as between every one of these combinations and its next recurrence every other possible combination would necessarily have been undergone, and since every one of these combinations would determine the whole series in the same order, a circular movement of absolutely identical series is thus demonstrated: the universe is thus shown to be a circular movement which has already repeated itself an infinite number of times, and which plays its game for all eternity.[§]

[†] See W. A. Kaufman, *Nietzsche: Philosopher, Psychologist, Antichrist*, Princeton University Press, 1950, Chapter 11.

[‡] See C. Andler, *Nietzsche, sa Vie et sa Pensée*, Gallimard, Paris, 1958, Vol. 4, Livre 2, Chap. I and Livre 3, Chap. I.

[§] F. Nietzsche, *Der Wille zur Macht*, in his *Gesammelte Werke*, Musarion Verlag, Munich, 1926, Vol. 19, Book 4, Part 3; English translation by O. Manthey-Zorn in Nietzsche, *An Anthology of his works*, Washington Square Press, New York, 1964, p. 90.

Nietzsche thought that his doctrine was not materialistic because materialism entailed the irreversible dissipation of energy and the ultimate heat death of the universe. In fact, the discussion of Poincaré's theorem (Selections 6–10) showed that on the contrary it is precisely the mechanistic view of the universe that has recurrence as its inevitable consequence. Since Zermelo and some other scientists believed that the second law must have absolute rather than merely statistical validity, they thought that the mechanistic theory was refuted by the " recurrence paradox ".† The effect of Nietzsche's argument is actually just the opposite of what he thought it should be: if there *is* eternal recurrence, so that the second law of thermodynamics cannot always be valid, then the materialist view (as represented by Boltzmann's interpretation) would be substantiated.

While Poincaré's version of the theorem, which was used by Zermelo to attack the kinetic theory (Selections 7 and 9), had an important influence on the history of theoretical physics whereas Nietzsche's had none, it should not be thought that the former was a rigorous mathematical proof whereas the latter was merely another aphorism of the mad philosopher. Nietzsche's version presents the essential point clearly and plausibly enough, while Poincaré and Zermelo, with all their concern for mathematical rigor, still fail to come up to the standards of modern mathematics. Their proof is inadequate because they lack the precise concept of the " measure " of a set of points, introduced by Lebesgue in 1902 and used to prove the recurrence theorem by Carathéodory in 1918.‡ Poincaré's " exceptional " non-recurrent trajectories, which can be proved to exist and are in fact infinite in

---

† See also F. Wald, *Die Energie und ihre Entwerthung*, Englemann, Leipzig, 1889, p. 104; E. Mach, *Die Prinzipien der Wärmelehre*, Barth, Leipzig, 1896, p. 362; G. Helm, *Die Lehre von der Energie historisch-kritisch entwickelt*, Felix, Leipzig, 1887; *Grundzüge der mathematischen Chemie*, Veit, Leipzig, 1898; P. Duhem, *Trauté d'energetique*, Gauthier-Villars, Paris, 1911; H. Poincaré, *Thermodynamique*, Gauthier-Villars, Paris, 1892; *Nature* **45**, 414, 485 (1892).

‡ H. Lebesgue, *Annali Mat. pura e appl.* **7**, 231 (1902); C. Carathéodory, *Sitzber. Preuss. Akad. Wiss.* 579 (1919).

number, yet have zero " probability ", are a " set of measure zero " with respect to the others, in modern terminology.

Who won the debate between Zermelo and Boltzmann? The reader may of course decide for himself, but modern physicists almost unanimously follow Boltzmann's views. To some extent this is because the alternative of rejecting all atomic theories is no longer open to us, as it was in the 1890's, and we are therefore forced to rely on something like an H-theorem in order to understand irreversibility. While physicists do not generally go so far as Reichenbach in accepting Boltzmann's notion of alternating time-directions in the universe,† they do accept the statistical interpretation of the second law of thermodynamics.‡

---

† H. Reichenbach, *The Direction of Time*. University of California Press, Berkeley, 1956.

‡ A concrete example of the behaviour of the H-curve postulated by Boltzmann was provided by the Ehrenfest urn-model: see P. and T. Ehrenfest, *Phys. z.* **8,** 311 (1907), and the book by ter Haar (1954).

Bibliography

BOLTZMANN, LUDWIG. *Vorlesungen über Gastheorie.* J. A. Barth, Leipzig, Part
I, 1896, Part II, 1898. English translation by S. G. Brush, *Lectures on Gas
Theory*, University of California Press, Berkeley, 1964.

BRUSH, STEPHEN G. Development of the Kinetic Theory of Gases, VI.
Viscosity. *Amer. J. Phys.* **30**, 269 (1962).

BRUSH, STEPHEN G. Thermodynamics and History, *The Graduate Journal* (in
press).

BRYAN, GEORGE H. Researches related to the connection of the second law
with dynamical principles, *Brit. Assoc. Rept.* **61**, 85 (1891).

BRYAN, GEORGE H. The laws of distribution of energy and their limitations,
Brit. Assoc. Rept. **64**, 64 (1894).

CARNOT, SADI. *Reflexions sur la puissance motrice de feu et sur les machines
propres à developper cette puissance.* Bachelier, Paris, 1824. Reprinted
with additional material by Blanchard, Paris, 1878. English translation
by R. H. Thurston, together with other papers by Clapeyron and Clausius,
in *Reflections on the Motive Power of Fire*, etc., Dover, New York, 1960.

CHAPMAN, SYDNEY. On the law of distribution of molecular velocities, and on
the theory of viscosity and thermal conduction, in a non-uniform simple
monatomic gas, *Phil. Trans. Roy. Soc. London*, **A216**, 279 (1916).

CHAPMAN, SYDNEY. On the kinetic theory of a gas, Part II, a composite monatomic gas, diffusion, viscosity, and thermal conduction, *Phil. Trans. Roy.
Soc. London*, **A217**, 115 (1917).

CHAPMAN, SYDNEY and COWLING, T. G. *The Mathematical Theory of Non-
Uniform Gases*, Cambridge University Press, 1939, 2nd edition, 1952.

DUGAS, RENÉ. *La Théorie Physique au sens de Boltzmann et ses prolongements
modernes.* Éditions du Griffon, Neuchâtel-Suisse, 1959.

EHRENFEST, PAUL and TATIANA. Begriffliche Grundlagen der statistischen
Auffassung in der Mechanik. *Encyklopädie der mathematischen Wissen-
schaften*, Vol. 4, Part 32, Teubner, Leipzig, 1911. English translation by
M. J. Moravcsik, *The Conceptual Foundations of the Statistical Approach
in Mechanics*, Cornell University Press, 1959.

ENSKOG, DAVID. *Kinetische Theorie der Vorgänge in mässig verdünnten Gasen*,
Almqvist and Wiksell, Uppsala, 1917.

FURRY, WENDELL H. On the elementary explanation of diffusion phenomena
in Gases, *Amer. J. Phys.* **16**, 63 (1948).

HAAR, D. TER. *Elements of Statistical Mechanics*, Rinehart, New York, 1954.

JEANS, JAMES H. *The Dynamical Theory of Gases*, Cambridge University Press,
1904; 4th edition, 1925, reprinted by Dover, New York, 1954.

19

KELVIN, WILLIAM THOMSON. On a universal tendency in nature to the dissipation of mechanical energy, *Phil Mag.* [4] **4**, 304 (1852); *Proc. Roy. Soc. Edinburgh* **3**, 139 (1857); reprinted in his *Mathematical and Physical Papers*, Cambridge University Press, 2nd edition, 1882–1911, Vol. 1. p. 511.

KENNARD, EARLE H. *Kinetic Theory of Gases with an Introduction to Statistical Mechanics*, McGraw-Hill, New York, 1938.

LEOB, LEONARD B. *Kinetic Theory of Gases*, McGraw-Hill, New York, 1922, 2nd edition, 1934.

TOLMAN, RICHARD C. *The Principles of Statistical Mechanics*, Oxford University Press, 1938.

Part 2

1

On the Dynamical Theory of Gases*

JAMES CLERK MAXWELL

SUMMARY†

The theory of transport processes in gases—such as diffusion, heat conduction, and viscosity—is developed on the basis of the assumption that the molecules behave like point-centres of force. The method of investigation consists in calculating mean values of various functions of the velocity of all the molecules of a given kind within an element of volume, and the variations of these mean values due, first, to the encounters of the molecules with others of the same or a different kind; second, to the action of external forces such as gravity; and third, to the passage of molecules through the boundary of the element of volume.

The encounters are analysed of molecules repelling each other with forces inversely as the nth power of the distance. In general the variation of mean values of functions of the velocity due to encounters depends on the relative velocity of the two colliding molecules, and unless the gas is in thermal equilibrium the velocity distribution is unknown so that these variations cannot be calculated directly. However, in the case of inverse fifth-power forces the relative velocity drops out, and the calculations can be carried out. It is found that in this special case the viscosity coefficient is proportional to the absolute temperature, in agreement with experimental results of the author. An expression for the diffusion coefficient is also derived, and compared with experimental results published by Graham.

A new derivation is given of the velocity–distribution law for a gas in thermal equilibrium. The theory is also applied to give an explanation of the

* Originally published in the *Philosophical Transactions of the Royal Society of London*, **157**, 49–88 (1867), and in *Philosophical Magazine*, **32**, 390–3 (1866), **35**, 129–45, 185–217 (1868); reprinted in *The Scientific Papers of James Clerk Maxwell*, Cambridge University Press, 1890, **2**, 26–78.

† All Summaries by S. G. B.

Law of Equivalent Volumes, the conduction of heat through gases, the hydro-dynamic equations of motion corrected for viscosity (Navier–Stokes equation), the relaxation of inequalities of pressure, and the final equilibrium of temperature in a column of gas under the influence of gravity.

Theories of the constitution of bodies suppose them either to be continuous and homogeneous, or to be composed of a finite number of distinct particles or molecules.

In certain applications of mathematics to physical questions, it is convenient to suppose bodies homogeneous in order to make the quantity of matter in each differential element a function of the co-ordinates, but I am not aware that any theory of this kind has been proposed to account for the different properties of bodies. Indeed the properties of a body supposed to be a uniform *plenum* may be affirmed dogmatically, but cannot be explained mathematically.

Molecular theories suppose that all bodies, even when they appear to our senses homogeneous, consist of a multitude of particles, or small parts the mechanical relations of which constitute the properties of the bodies. Those theories which suppose that the molecules are at rest relative to the body may be called statical theories, and those which suppose the molecules to be in motion, even while the body is apparently at rest, may be called dynamical theories.

If we adopt a statical theory, and suppose the molecules of a body kept at rest in their positions of equilibrium by the action of forces in the directions of the lines joining their centres, we may determine the mechanical properties of a body so constructed, if distorted so that the displacement of each molecule is a function of its co-ordinates when in equilibrium. It appears from the mathematical theory of bodies of this kind, that the forces called into play by a small change of form must always bear a fixed proportion to those excited by a small change of volume.

Now we know that in fluids the elasticity of form is evanescent, while that of volume is considerable. Hence such theories will not apply to fluids. In solid bodies the elasticity of form appears in many cases to be smaller in proportion to that of volume than the theory gives,† so that we are forced to give up the theory of molecules whose displacements are functions of their co-ordinates when at rest, even in the case of solid bodies.

The theory of moving molecules, on the other hand, is not open to these objections. The mathematical difficulties in applying the theory are considerable, and till they are surmounted we cannot fully decide on the applicability of the theory. We are able, however, to explain a great variety of phenomena by the dynamical theory which have not been hitherto explained otherwise.

The dynamical theory supposes that the molecules of solid bodies oscillate about their positions of equilibrium, but do not travel from one position to another in the body. In fluids the molecules are supposed to be constantly moving into new relative positions, so that the same molecule may travel from one part of the fluid to any other part. In liquids the molecules are supposed to be always under the action of the forces due to neighbouring molecules throughout their course, but in gases the greater part of the path of each molecule is supposed to be sensibly rectilinear and beyond the sphere of sensible action of the neighbouring molecules.

I propose in this paper to apply this theory to the explanation of various properties of gases, and to shew that, besides accounting for the relations of pressure, density, and temperature in a single gas, it affords a mechanical explanation of the known chemical relation between the density of a gas and its equivalent weight, commonly called the Law of Equivalent Volumes. It also explains the diffusion of one gas through another, the internal friction of a gas, and the conduction of heat through gases.

The opinion that the observed properties of visible bodies apparently at rest are due to the action of invisible molecules in rapid motion is to be found in Lucretius. In the exposition which he gives of the theories of Democritus as modified by Epicurus, he

---

† In glass, according to Dr. Everett's second series of experiments (1866), the ratio of the elasticity of form to that of volume is greater than that given by the theory. In brass and steel it is less. March 7, 1867. [J. D. Everett, *Phil. Trans.* **156**, 185 (1866).]

describes the invisible atoms as all moving downwards with equal velocities, which, at quite uncertain times and places, suffer an imperceptible change, just enough to allow of occasional collisions taking place between the atoms. These atoms he supposes to set small bodies in motion by an action of which we may form some conception by looking at the motes in a sunbeam. The language of Lucretius must of course be interpreted according to the ideas of his age, but we need not wonder that it suggested to Le Sage the fundamental conception of his theory of gases, as well as his doctrine of ultramundane corpuscles.†

Professor Clausius, to whom we owe the most extensive developments of dynamical theory of gases, has given‡ a list of authors who have adopted or given countenance to any theory of invisible particles in motion. Of these, Daniel Bernoulli, in the tenth section of his *Hydrodynamics*, distinctly explains the pressure of air by the impact of its particles on the sides of the vessel containing it.§

Clausius also mentions a book entitled *Deux Traités de Physique Mécanique*, publiés par Pierre Prevost, comme simple Éditeur du premier et comme Auteur du second, Genève et Paris, 1818. The first memoir is by G. Le Sage, who explains gravity by the impact of " ultramundane corpuscles " on bodies. These corpuscles also set in motion the particles of light and various ethereal media, which in their turn act on the molecules of gases and keep up their motions. His theory of impact is faulty, but his explanation of the expansive force of gases is essentially the same as in the dynamical theory as it now stands. The second memoir, by Prevost, contains new applications of the principles of Le Sage to gases and to light. A more extensive application of the theory of

---

† [G. L. LeSage, *Physique Mecanique* (1746), published by P. Prevost under the title *Deux Traites de Physique Mecanique*, Geneva, 1818; see also LeSage's article " Lucrèce Newtonien " in *Nouveaux Mémoires de l'Academie Royale des Sciences et Belles-Lettres*, Berlin, 1782, p. 404.]

‡ Poggendorff's *Annalen*, January 1862. Translated by G. C. Foster, B.A., *Phil. Mag.* June 1862. [*Ann. Phys.* **115**, 1; *Phil. Mag.* **23**, 417, 512.]

[See Vol. 1 of this series, Selection 3.]

moving molecules was made by Herapath.† His theory of the collisions of perfectly hard bodies, such as he supposes the molecules to be, is faulty, inasmuch as it makes the result of impact depend on the absolute motion of the bodies, so that by experiments on such hard bodies (if we could get them) we might determine the absolute direction and velocity of the motion of the earth.‡ This author, however, has applied his theory to the numerical results of experiment in many cases, and his speculations are always ingenious, and often throw much real light on the questions treated. In particular, the theory of temperature and pressure in gases and the theory of diffusion are clearly pointed out.

Dr Joule§ has also explained the pressure of gases by the impact of their molecules, and has calculated the velocity which they must have in order to produce the pressure observed in particular gases.

It is to Professor Clausius, of Zurich, that we owe the most complete dynamical theory of gases. His other researches on the general dynamical theory of heat are well known, and his memoirs *On the kind of Motion which we call Heat*, are a complete exposition of the molecular theory adopted in this paper. After reading his investigation‖ of the distance described by each molecule between successive collisions, I published some propositions¶ on the motions and collisions of perfectly elastic spheres, and deduced several properties of gases, especially the law of equivalent volumes, and the nature of gaseous friction. I also gave a theory of diffusion of gases, which I now know to be erroneous, and there were several errors in my theory of the conduction of heat in gases

---

† *Mathematical Physics*, etc., by John Herapath, Esq. 2 vols. London: Whittaker and Co., and Herapath's *Railway Journal* Office, 1847. [See also *Annals of Philosophy* [2] **1**, 273, 340, 401; **2**, 50, 89, 201, 257, 363, 435; **3**, 16 (1821).]

‡ *Mathematical Physics*, etc., p. 134.

§ *Some Remarks on Heat and the Constitution of Elastic Fluids*, October 3, 1848. *Memoirs of the Manchester Literary and Philosophical Society* **9**, 107 (1851).

‖ *Phil. Mag.* February 1859. [Part A, Selection 9.]

¶ " Illustrations of the Dynamical Theory of Gases," *Phil. Mag.* January and July 1860. [Vol. 1, Selection 10.]

which M. Clausius has pointed out in an elaborate memoir on that subject.†

M. O. E. Meyer‡ has also investigated the theory of internal friction on the hypothesis of hard elastic molecules.

In the present paper I propose to consider the molecules of a gas, not as elastic spheres of definite radius, but as small bodies or groups of smaller molecules repelling one another with a force whose direction always passes very nearly through the centres of gravity of the molecules, and whose magnitude is represented very nearly by some function of the distance of the centres of gravity. I have made this modification of the theory in consequence of the results of my experiments on the viscosity of air at different temperatures, and I have deduced from these experiments that the repulsion is inversely as the *fifth* power of the distance.

If we suppose an imaginary plane drawn through a vessel containing a great number of such molecules in motion, then a great many molecules will cross the plane in either direction. The excess of the mass of those which traverse the plane in the positive direction over that of those which traverse it in the negative direction, gives a measure of the flow of gas through the plane in in the positive direction.

If the plane be made to move with such a velocity that there is no excess of flow of molecules in one direction through it, then the velocity of the plane is the mean velocity of the gas resolved normal to the plane.

There will still be molecules moving in both directions through the plane, and carrying with them a certain amount of momentum into the portion of gas which lies on the other side of the plane.

The quantity of momentum thus communicated to the gas on the other side of the plane during a unit of time is a measure of the force exerted on this gas by the rest. This force is called the pressure of the gas.

---

† Poggendorff, January 1862; *Phil Mag.* June 1862. Translated by G. C. Foster, B.A., *Phil. Mag.* June 1862. [*Ann. Phys.* **115**, 1; *Phil. Mag.* **23**, 417, 512.]

‡ " Ueber die innere Reibung der Gase " (Poggendorff, Vol. CXXV. 1865).

If the velocities of the molecules moving in different directions were independent of one another, then the pressure at any point of the gas need not be the same in all directions, and the pressure between two portions of gas separated by a plane need not be perpendicular to that plane. Hence, to account for the observed equality of pressure in all directions, we must suppose some cause equalizing the motion in all directions. This we find in the deflection of the path of one particle by another when they come near one another. Since, however, this equalization of motion is not instantaneous, the pressures in all directions are perfectly equalized only in the case of a gas at rest, but when the gas is in a state of motion, the want of perfect equality in the pressures gives rise to the phenomena of viscosity or internal friction. The phenomena of viscosity in all bodies may be described, independently of hypothesis, as follows:

A distortion or strain of some kind, which we may call S, is produced in the body by displacement. A state of stress or elastic force which we may call F is thus excited. The relation between the stress and the strain may be written $F = ES$, where E is the coefficient of elasticity for that particular kind of strain. In a solid body free from viscosity, F will remain $= ES$, and

$$\frac{dF}{dt} = E\frac{dS}{dt}$$

If, however, the body is viscous, F will not remain constant, but will tend to disappear at a rate depending on the value of F, and on the nature of the body. If we suppose this rate proportional to F, the equation may be written

$$\frac{dF}{dt} = E\frac{dS}{dt} - \frac{F}{T}$$

which will indicate the actual phenomena in an empirical manner. For if S be constant,

$$F = ESe^{-t/T}$$

showing that F gradually disappears, so that if the body is left to

B*

itself it gradually loses any internal stress, and the pressures are finally distributed as in a fluid at rest.

If dS/dl is constant, that is, if there is a steady motion of the body which continually increases the displacement,

$$F = ET\frac{dS}{dt} + Ce^{-t/T}$$

showing that F tends to a constant value depending on the rate of displacement. The quantity ET, by which the rate of displacement must be multiplied to get the force, may be called the coefficient of viscosity. It is the product of a coefficient of elasticity, E, and a time T, which may be called the " time of relaxation " of the elastic force. In mobile fluids T is a very small fraction of a second, and E is not easily determined experimentally. In viscous solids T may be several hours or days, and then E is easily measured. It is possible that in some bodies T may be a function of F, and this would account for the gradual untwisting of wires after being twisted beyond the limit of perfect elasticity. For if T diminishes as F increases, the parts of the wire furthest from the axis will yield more rapidly than the parts near the axis during the twisting process, and when the twisting force is removed, the wire will at first untwist till there is equilibrium between the stresses in the inner and outer portions. These stresses will then undergo a gradual relaxation; but since the actual value of the stress is greater in the outer layers, it will have a more rapid rate of relaxation, so that the wire will go on gradually untwisting for some hours or days, owing to the stress on the interior portions maintaining itself longer than that of the outer parts. This phenomenon was observed by Weber in silk fibres, by Kohlrausch in glass fibres, and by myself in steel wires.

In the case of a collection of moving molecules such as we suppose a gas to be, there is also a resistance to change of form, constituting what may be called the linear elasticity, or " rigidity " of the gas, but this resistance gives way and diminishes at a rate depending on the amount of the force and on the nature of the gas.

Suppose the molecules to be confined in a rectangular vessel with perfectly elastic sides, and that they have no action on one another, so that they never strike one another, or cause each other to deviate from their rectilinear paths. Then it can easily be shewn that the pressures on the sides of the vessel due to the impacts of the molecules are perfectly independent of each other, so that the mass of moving molecules will behave, not like a fluid, but like an elastic solid. Now suppose the pressures at first equal in the three directions perpendicular to the sides, and let the dimensions a, b, c of the vessel be altered by small quantities, δa, δb, δc.

Then if the original pressure in the direction of a was p, it will become

$$p\left(1-3\frac{\delta a}{a}-\frac{\delta b}{b}-\frac{\delta c}{c}\right);$$

or if there is no change of volume,

$$\frac{\delta p}{p}=-2\frac{\delta a}{a},$$

shewing that in this case there is a " longitudinal " elasticity of form of which the coefficient is $2p$. The coefficient of " Rigidity " is therefore $= p$.

This rigidity, however, cannot be directly observed, because the molecules continually deflect each other from their rectilinear courses, and so equalize the pressure in all directions. The rate at which this equalization takes place is great, but not infinite; and therefore there remains a certain inequality of pressure which constitutes the phenomenon of viscosity.

I have found by experiment that the coefficient of viscosity in a given gas is independent of the density, and proportional to the absolute temperature, so that if ET be the viscosity, $ET \propto p/\rho$.

But $E = p$, therefore T, the time of relaxation, varies inversely as the density and is independent of the temperature. Hence, the number of collisions producing a given deflection which take place in unit of time is independent of the temperature, that is, of the velocity of the molecules, and is proportional to the number of molecules in unit of volume. If we suppose the molecules hard elastic bodies, the number of collisions of a given kind will be proportional to the velocity, but if we suppose them centres of force, the angle of deflection will be smaller when the velocity is greater; and if the force is inversely as the fifth power of the distance, the number of deflections of a given kind will be independent of the velocity. Hence I have adopted this law in making my calculations.

The effect of the mutual action of the molecules is not only to equalize the pressure in all directions, but, when molecules of different kinds are present, to communicate motion from the one kind to the other. I formerly shewed that the final result in the case of hard elastic bodies is to cause the average *vis viva* of a molecule to be the same for all the different kinds of molecules. Now the pressure due to each molecule is proportional to its *vis viva*, hence the whole pressure due to a given number of molecules in a given volume will be the same whatever the mass of the molecules, provided the molecules of different kinds are permitted freely to communicate motion to each other.

When the flow of *vis viva* from the one kind of molecules to the other is zero, the temperature is said to be the same. Hence, equal volumes of different gases at equal pressures and temperatures contain equal numbers of molecules.

This result of the dynamical theory affords the explanation of the " law of equivalent volumes " in gases.

We shall see that this result is true in the case of molecules acting as centres of force. A law of the same general character is probably to be found connecting the temperatures of liquid and solid bodies with the energy possessed by their molecules, although our ignorance of the nature of the connexions between the molecules renders it difficult to enunciate the precise form of the law.

The molecules of a gas in this theory are those portions of it which move about as a single body. These molecules may be mere points, or pure centres of force endowed with inertia, or the capacity of performing work while losing velocity. They may be systems of several such centres of force, bound together by their mutual actions, and in this case the different centres may either be separated, so as to form a group of points, or they may be actually coincident, so as to form one point.

Finally, if necessary, we may suppose them to be small solid bodies of a determinate form; but in this case we must assume a new set of forces binding the parts of these small bodies together, and so introduce a molecular theory of the second order. The doctrines that all matter is extended, and that no two portions of matter can coincide in the same place, being deductions from our experiments with bodies sensible to us, have no application to the theory of molecules.

The actual energy of a moving body consists of two parts, one due to the motion of its centre of gravity, and the other due to the motions of its parts relative to the centre of gravity. If the body is of invariable form, the motions of its parts relative to the centre of gravity consist entirely of rotation, but if the parts of the body are not rigidly connected, their motions may consist of oscillations of various kinds, as well as rotation of the whole body.

The mutual interference of the molecules in their courses will cause their energy of motion to be distributed in a certain ratio between that due to the motion of the centre of gravity and that due to the rotation, or other internal motion. If the molecules are pure centres of force, there can be no energy of rotation, and the whole energy is reduced to that of translation; but in all other cases the whole energy of the molecule may be represented by $\frac{1}{2}Mv^2\beta$, where β is the ratio of the total energy to the energy of translation. The ratio β will be different for every molecule, and will be different for the same molecule after every encounter with another molecule, but it will have an average value depending on the nature of the molecules, as has been shown by Clausius.† The value of β can be determined if we know either of the specific heats of the gas, or the ratio between them.

The method of investigation which I shall adopt in the following paper, is to determine the mean values of the following functions of the velocity of all the molecules of a given kind within an element of volume:

---

† [Vol. 1, Selection 8.]

(α) the mean velocity resolved parallel to each of the coordinate axes;

(β) the mean values of functions of two dimensions of these component velocities;

(γ) The mean values of functions of three dimensions of these velocities.

The rate of translation of the gas, whether by itself, or by diffusion through another gas, is given by (α), the pressure of the gas on any plane, whether normal or tangential to the plane, is given by (β), and the rate of conduction of heat through the gas is given by (γ).

I propose to determine the variations of these quantities, due, 1st, to the encounters of the molecules with others of the same system or of a different system; 2nd, to the action of external forces such as gravity; and 3rd, to the passage of molecules through the boundary of the element of volume.

I shall then apply these calculations to the determination of the statical cases of the final distribution of two gases under the action of gravity, the equilibrium of temperature between two gases, and the distribution of temperature in a vertical column. These results are independent of the law of force between the molecules. I shall also consider the dynamical cases of diffusion, viscosity, and conduction of heat, which involve the law of force between the molecules.

On the Mutual Action of Two Molecules

Let the masses of these molecules be M_1, M_2, and let their velocities resolved in three directions at right angles to each other be ξ_1, η_1, ζ_1 and ξ_2, η_2, ζ_2. The components of the velocity of the centre of gravity of the two molecules will be

$$\frac{\xi_1 M_1 + \xi_2 M_2}{M_1 + M_2}, \quad \frac{\eta_1 M_1 + \eta_2 M_2}{M_1 + M_2}, \quad \frac{\zeta_1 M_1 + \zeta_2 M_2}{M_1 + M_2}$$

The motion of the centre of gravity will not be altered by the mutual action of the molecules, of whatever nature that action may be. We may therefore take the centre of gravity as the origin of a system of coordinates moving parallel to itself with uniform velocity, and consider the alteration of the motion of each particle with reference to this point as origin.

If we regard the molecules as simple centres of force, then each molecule will describe a plane curve about this centre of gravity, and the two curves will be similar to each other and symmetrical with respect to the line of apses. If the molecules move with sufficient velocity to carry them out of the sphere of their mutual action, their orbits will each have a pair of asymptotes inclined at an angle $\pi/2 - \theta$ to the line of apses. The asymptotes of the orbit of M_1 will be at a distance b_1 from the centre of gravity, and those of M_2 at a distance b_2, where

$$M_1 b_1 = M_2 b_2$$

The distance between two parallel asymptotes, one in each orbit, will be

$$b = b_1 + b_2$$

If, while the two molecules are still beyond each other's action, we draw a straight line through M_1 in the direction of the relative velocity of M_1 to M_2, and draw from M_2 a perpendicular to this line, the length of this perpendicular will be b, and the plane including b and the direction of relative motion will be the plane of the orbits about the centre of gravity.

When, after their mutual action and deflection, the molecules have again reached a distance such that there is no sensible action between them, each will be moving with the same velocity relative to the centre of gravity that it had before the mutual action, but the direction of this relative velocity will be turned through an angle 2θ in the plane of the orbit.

The angle θ is a function of the relative velocity of the molecules and of b, the form of the function depending on the nature of the action between the molecules.

If we suppose the molecules to be bodies, or systems of bodies, capable of rotation, internal vibration, or any form of energy other than simple motion of translation, these results will be modified. The value of θ and the final velocities of the molecules will depend on the amount of internal energy in each molecule before the encounter, and on the particular form of that energy at every instant during the mutual action. We have no means of determining such intricate actions in the present state of our knowledge of molecules, so that we must content ourselves with the assumption that the value of θ is, on an average, the same as for pure centres of force, and that the final velocities differ from the initial velocities only by quantities which may in each collision be neglected, although in a great many encounters the energy of translation and the internal energy of the molecules arrive, by repeated small exchanges, at a final ratio, which we shall suppose to be that of 1 to $\beta - 1$.

We may now determine the final velocity of M_1 after it has passed beyond the sphere of mutual action between itself and M_2.

Let V be the velocity of M_1 relative to M_2, then the components of V are

$$\xi_1 - \xi_2, \quad \eta_1 - \eta_2, \quad \zeta_1 - \zeta_2$$

The plane of the orbit is that containing V and b. Let this plane be inclined ϕ to a plane containing V and parallel to the axis of x; then, since the direction of V is turned round an angle 2θ in the plane of the orbit, while its magnitude remains the same, we may find the value of ξ_1 after the encounter. Calling it ξ_1',

$$\left. \begin{aligned} \xi_1' = \xi_1 + \frac{M_2}{M_1 + M_2} \{ (\xi_2 - \xi_1) 2 \sin^2 \theta \\ + \sqrt{(\eta_2 - \eta_1)^2 + (\zeta_2 - \zeta_1)^2} \sin 2\theta \cos \phi \} \end{aligned} \right\} \tag{1}$$

There will be similar expressions for the components of the final velocity of M_1 in the other coordinate directions.

If we know the initial positions and velocities of M_1 and M_2, we can determine V, the velocity of M_1 relative to M_2; b the shortest distance between M_1 and M_2 if they had continued to move with uniform velocity in straight lines; and ϕ the angle which determines the plane in which V and b lie. From V and b we can determine θ, if we know the law of force, so that the problem is solved in the case of two molecules.

When we pass from this case to that of two systems of moving molecules, we shall suppose that the time during which a molecule is beyond the action of other molecules is so great compared with the time during which it is deflected by that action, that we may neglect both the time and distance described by the molecules during the encounter, as compared with the time and the distance described while the molecules are free from disturbing force. We may also neglect those cases in which three or more molecules are within each other's spheres of action at the same instant.

On the Mutual Action of Two Systems of Moving Molecules

Let the number of molecules of the first kind in unit of volume be N_1, the mass of each being M_1. The velocities of these molecules will in general be different both in magnitude and direction. Let us select those molecules the components of whose velocities lie between

$$\xi_1 \text{ and } \xi_1 + d\xi_1, \quad \eta_1 \text{ and } \eta_1 + d\eta_1, \quad \zeta_1 \text{ and } \zeta_1 + d\zeta_1$$

and let the number of these molecules be dN_1. The velocities of these molecules will be very nearly equal and parallel.

On account of the mutual actions of the molecules, the number of molecules which at a given instant have velocities within given limits will be definite, so that

$$dN_1 = f_1(\xi_1\eta_1\zeta_1)\,d\xi_1 d\eta_1 d\zeta_1 \tag{2}$$

We shall consider the form of this function afterwards.

Let the number of molecules of the second kind in unit of volume be N_2, and let dN_2 of these have velocities between ξ_2 and $\xi_2 + d\xi_2$, η_2 and $\eta_2 + d\eta_2$, ζ_2 and $\zeta_2 + d\zeta_2$, where

$$dN_2 = f_2(\xi_2\eta_2\zeta_2)\,d\xi_2 d\eta_2 d\zeta_2$$

The velocity of any of the dN_1 molecules of the first system relative to the dN_2 molecules of the second system is V, and each molecule M_1 will in the time δt describe a relative path $V\delta t$ among the molecules of the second system. Conceive a space bounded by the following surfaces. Let two cylindrical surfaces have the common axis $V\delta t$ and radii b and $b+db$. Let two planes be drawn through the extremities of the line $V\delta t$ perpendicular to it. Finally, let two planes be drawn through $V\delta t$ making angles ϕ and $\phi+d\phi$ with a plane through V parallel to the axis of x. Then the volume included between the four planes and the two cylindric surfaces will be $Vbdbd\phi\delta t$.

If this volume includes one of the molecules M_2, then during the time δt there will be an encounter between M_1 and M_2, in which b is between b and $b+db$, and ϕ between ϕ and $\phi+d\phi$.

Since there are dN_1 molecules similar to M_1 and dN_2 similar to M_2 in unit of volume, the whole number of encounters of the given kind between the two systems will be

$$V b\,db\,d\phi\,\delta t\,dN_1 dN_2$$

Now let Q be any property of the molecule M_1, such as its velocity in a given direction, the square or cube of that velocity or any other property of the molecule which is altered in a known manner by an encounter of the given kind, so that Q becomes Q' after the encounter, then during the time δt a certain number of the molecules of the first kind have Q changed to Q', while the remainder retain the original value of Q, so that

$$\delta Q dN_1 = (Q'-Q)V b\,db\,d\phi\,\delta t\,dN_1 dN_2$$

or

$$\frac{\delta Q dN_1}{\delta t} = (Q'-Q)V b\,db\,d\phi\,dN_1 dN_2 \tag{3}$$

Here $\delta Q dN_1/\delta t$ refers to the alteration in the sum of the values of Q for the dN_1 molecules, due to their encounters of the given kind with the dN_2 molecules of the second sort. In order to determine the value of $\delta Q N_1/\delta t$, the rate of alteration of Q among all the molecules of the first kind, we must perform the following integrations:

1st, with respect to ϕ from $\phi = 0$ to $\phi = 2\pi$.

2nd, with respect to b from $b = 0$ to $b = \infty$. These operations will give the results of the encounters of every kind between the dN_1 and dN_2 molecules.

3rd, with respect to dN_2, or $f_2(\xi_2 \eta_2 \zeta_2)d\xi_2 d\eta_2 d\zeta_2$.

4th, with respect to dN_1, or $f_1(\xi_1 \eta_1 \zeta_1)d\xi_1 d\eta_1 d\zeta_1$.

These operations require in general a knowledge of the forms of f_1 and f_2.

1st. *Integration with respect to ϕ*

Since the action between the molecules is the same in whatever plane it takes place, we shall first   determine the value of $\int_0^{2\pi}(Q'-Q)d\phi$ in several cases, making Q some function of ξ, η, and ζ.

(α) Let $Q = \xi_1$ and $Q' = \xi_1'$, then

$$\int_0^{2\pi} (\xi_1'-\xi_1)d\phi = \frac{M_2}{M_1+M_2}(\xi_2-\xi_1)4\pi \sin^2 \theta \qquad (4)$$

(β) Let $Q = \xi_1^2$ and $Q' = \xi_1'^2$

$$\left.\begin{array}{l} \int_0^{2\pi} (\xi_1'^2-\xi_1^2)d\phi = \dfrac{M_2}{(M_1+M_2)^2}[(\xi_2-\xi_1)(M_1\xi_1 + M_2\xi_2) \\[2mm] \qquad \times 8\pi \sin^2 \theta + M_2\{(\eta_2-\eta_1)^2+(\zeta_2-\zeta_1)^2 \\[2mm] \qquad\qquad -2(\xi_2-\xi_1)^2\}\pi \sin^2 2\theta] \end{array}\right\} \qquad (5)$$

By transformation of coordinates we may derive from this

$$
\left.\begin{aligned}
\int_0^\pi (\xi_1' \eta_1' - \xi_1 \eta_1) d\phi = \frac{M_2}{(M_1 + M_2)^2} \big[&\{ M_2 \xi_2 \eta_2 - M_1 \xi_1 \eta_1 \\
&+ \tfrac{1}{2}(M_1 - M_2)(\xi_1 \eta_2 + \xi_2 \eta_1) \} \, 8\pi \sin^2\theta \\
&- 3M_2(\xi_2 - \xi_1)(\eta_2 - \eta_1) \big]
\end{aligned}\right\} \quad (6)
$$

with similar expressions for the other quadratic functions of ξ, η, ζ.

(γ) Let $Q = \xi_1(\xi_1^2 + \eta_1^2 + \zeta_1^2)$, and $Q' = \xi_1'(\xi_1'^2 + \eta_1'^2 + \zeta_1'^2)$;

then putting

$$
\xi_1^2 + \eta_1^2 + \zeta_1^2 = V_1^2, \quad \xi_1 \xi_2 + \eta_1 \eta_2 + \zeta_1 \zeta_2 = U, \quad \xi_2^2 + \eta_2^2 + \zeta_2^2 = V_2^2,
$$

and

$$
(\xi_2 - \xi_1)^2 + (\eta_2 - \eta_1)^2 + (\zeta_2 - \zeta_1)^2 = V^2
$$

we find

$$
\left.\begin{aligned}
\int_0^\pi (\xi_1' V_1'^2 - \xi_1 V_1^2) d\phi = &\frac{M_2}{M_1 + M_2} 4\pi \sin^2\theta \{ (\xi_2 - \xi_1) V_1^2 \\
&\qquad\qquad\qquad\qquad + 2\xi_1(U - V_1^2) \} \\
&+ \left(\frac{M_2}{M_1 + M_2}\right)^2 (8\pi \sin^2\theta - 3\pi \sin^2 2\theta) \times 2(\xi_2 - \xi_1)(U - V_1^2) \\
&+ \left(\frac{M_2}{M_1 + M_2}\right)^2 (8\pi \sin^2\theta + 2\pi \sin^2 2\theta) \times \xi_1 V^2 \\
&+ \left(\frac{M_2}{M_1 + M_2}\right)^2 (8\pi \sin^2\theta - 2\pi \sin^2 2\theta) \times 2(\xi_2 - \xi_1) V^2
\end{aligned}\right\} \quad (7)
$$

These are the principal functions of ξ, η, ζ whose changes we shall have to consider; we shall indicate them by the symbols α, β, or γ, according as the function of the velocity is of one, two, or three dimensions.

2nd. *Integration with respect to b*

We have next to multiply these expressions by bdb, and to integrate with respect to b from $b = 0$ to $b = \infty$. We must bear in mind that θ is a function of b and V, and can only be determined when the law of force is known. In the expressions which we have to deal with, θ occurs under two forms only, namely, $\sin^2\theta$ and $\sin^2 2\theta$. If, therefore, we can find the two values of

$$B_1 = \int_0^\infty 4\pi bdb \sin^2\theta, \quad \text{and} \quad B_2 = \int_0^\infty \pi bdb \sin^2 2\theta \tag{8}$$

we can integrate all the expressions with respect to b.

B_1 and B_2 will be functions of V only, the form of which we can determine only in particular cases, after we have found θ as a function of b and V.

Determination of θ for Certain Laws of Force

Let us assume that the force between the molecules M_1 and M_2 is repulsive and varies inversely as the nth power of the distance between them, the value of the moving force at distance unity being K, then we find by the equation of central orbits,

$$\frac{\pi}{2} - \theta = \int_0^{x'} \frac{dx}{\sqrt{1 - x^2 - \dfrac{2}{n-1}\left(\dfrac{x}{a}\right)^{n-1}}} \tag{9}$$

where $x = b/r$, or the ratio of b to the distance of the molecules at a given time: x is therefore a numerical quantity; α is also a numerical quantity and is given by the equation

$$\alpha = b \left\{ \frac{V^2 M_1 M_2}{K(M_1 + M_2)} \right\}^{1/(n-1)} \tag{10}$$

The limits of integration are $x = 0$ and $x = x'$, where x' is the least positive root of the equation

$$1 - x^2 - \frac{2}{n-1}\left(\frac{x}{\alpha}\right)^{n-1} = 0 \qquad (11)$$

It is evident that θ is a function of α and n, and when n is known θ may be expressed as a function of α only.

Also

$$bdb = \left\{\frac{K(M_1+M_2)}{V^2 M_1 M_2}\right\}^{2/(n-1)} \alpha d\alpha \qquad (12)$$

so that if we put

$$A_1 = \int\limits_0^\infty 4\pi\alpha d\alpha \sin^2\theta, \quad A_2 = \int\limits_0^\infty \pi\alpha d\alpha \sin^2 2\theta \qquad (13)$$

A_1 and A_2 will be definite numerical quantities which may be ascertained when n is given, and B_1 and B_2 may be found by multiplying A_1 and A_2 by

$$\left\{\frac{K(M_1+M_2)}{M_1 M_2}\right\}^{2/(n-1)} V^{-4/(n-1)}$$

Before integrating further we have to multiply by V, so that the form in which V will enter into the expressions which have to be integrated with respect to dN_1 and dN_2 will be

$$V^{(n-5)/(n-1)}$$

It will be shewn that we have reason from experiments on the viscosity of gases to believe that $n = 5$. In this case V will disappear from the expressions of the form (3), and they will be capable of immediate integration with respect to dN_1 and dN_2.

If we assume $n = 5$ and put $\alpha^4 = 2\cot^2 2\phi$ and

$$x = \sqrt{1 - \tan^2\phi}\cos\psi,$$

$$\left.\begin{aligned}\frac{\pi}{2} - \theta &= \sqrt{\cos 2\phi} \int\limits_0^{\pi/2} \frac{d\psi}{\sqrt{1 - \sin^2\phi\sin^2\psi}} \\ &= \sqrt{\cos 2\phi}\, F_{\sin\phi}\end{aligned}\right\} \qquad (14)$$

where $F_{\sin \phi}$ is the complete elliptic function of the first kind and is given in Legendre's Tables. I have computed the following Table of the distance of the asymptotes, the distance of the apse, the value of θ, and of the quantities whose summation leads to A_1 and A_2.

ϕ		b	Distance of apse	θ		$\dfrac{\sin^2 \theta}{\sin^2 2\phi}$	$\dfrac{\sin^2 2\theta}{\sin^2 2\phi}$
°	′			°	′		
0	0	infinite	infinite	0	0	0	0
5	0	2381	2391	0	31	·00270	·01079
10	0	1658	1684	1	53	·01464	·03689
15	0	1316	1366	4	47	·02781	·11048
20	0	1092	1172	8	45	·05601	·21885
25	0	916	1036	14	15	·10325	·38799
30	0	760	931	21	42	·18228	·62942
35	0	603	845	31	59	·31772	·71433
40	0	420	772	47	20	·55749	1·02427
41	0	374	758	51	32	·62515	·96763
42	0	324	745	56	26	·70197	·85838
43	0	264	732	62	22	·78872	·67868
44	0	187	719	70	18	·88745	·40338
44	30	132	713	76	1	·94190	·21999
45	0	0	707	90	0	1·00000	·00000

$$A_1 = \int 4\pi \alpha d\alpha \sin^2 \theta = 2\cdot6595 \tag{15}$$

$$A_2 = \int \pi \alpha d\alpha \sin^2 2\theta = 1\cdot3682 \tag{16}$$

The paths described by molecules about a centre of force S, repelling inversely as the fifth power of the distance, are given in Fig. 1 overleaf.

The molecules are supposed to be originally moving with equal velocities in parallel paths, and the way in which their deflections depend on the distance of the path from S is shewn by the different curves in the figure.

3rd. *Integration with respect to* dN_2

We have now to integrate expressions involving various functions of ξ, η, ζ, and V with respect to all the molecules of the second sort. We may write the expression to be integrated

$$\iiint Q V^{(n-5)/(n-1)} f_2(\xi_2 \eta_2 \zeta_2) d\xi_2 d\eta_2 d\zeta_2$$

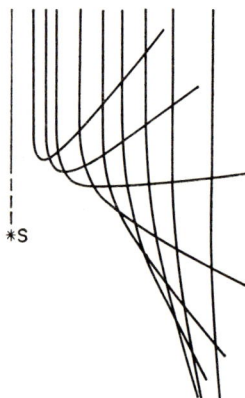

Fig. 1.

where Q is some function of ξ, η, ζ, etc., already determined, and f_2 is the function which indicates the distribution of velocity among the molecules of the second kind.

In the case in which $n = 5$, V disappears, and we may write the result of integration

$$\bar{Q} N_2,$$

where $\bar{Q}$ is the mean value of Q for all the molecules of the second kind, and N_1 is the number of those molecules.

If, however, n is not equal to 5, so that V does not disappear, we should require to know the form of the function f_2 before we could proceed further with the integration.

The only case in which I have determined the form of this function is that of one or more kinds of molecules which have by their continual encounters brought about a distribution of velocity such that the number of molecules whose velocity lies within given limits remains constant. In the *Philosophical Magazine* for January 1860, I have given an investigation of this case, founded on the assumption that the probability of a molecule having a velocity resolved parallel to x lying between given limits is not in any way affected by the knowledge that the molecule has a given velocity resolved parallel to y.† As this assumption may appear precarious, I shall now determine the form of the function in a different manner.

On the Final Distribution of Velocity among the Molecules of Two Systems acting on one another according to any Law of Force

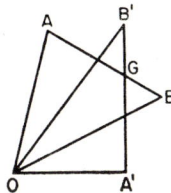

FIG. 2.

From a given point O let lines be drawn representing in direction and magnitude the velocities of every molecule of either kind in unit of volume. The extremities of these lines will be distributed over space in such a way that if an element of volume dV be taken anywhere, the number of such lines which will terminate within dV will be $f(r)dV$, where r is the distance of dV from O.

Let $OA = a$ be the velocity of a molecule of the first kind, and $OB = b$ that of a molecule of the second kind before they encounter one another, then BA will be the velocity of A relative

† [Vol. 1, Selection 10.]

to B; and if we divide AB in G inversely as the masses of the molecules, and join OG, OG will be the velocity of the centre of gravity of the two molecules.

Now let $OA' = a'$ and $OB' = b'$ be the velocities of the two molecules after the encounter, $GA = GA'$ and $GB = GB'$, and $A'GB'$ is a straight line not necessarily in the plane of OAB. Also $AGA' = 2\theta$ is the angle through which the relative velocity is turned in the encounter in question. The relative motion of the molecules is completely defined if we know BA the relative velocity before the encounter, 2θ the angle through which BA is turned during the encounter, and ϕ the angle which defines the direction of the plane in which BA and $B'A'$ lie. All encounters in which the magnitude and direction of BA, and also θ and ϕ, lie within certain almost contiguous limits, we shall class as encounters of the given kind. The number of such encounters in unit of time will be

$$n_1 n_2 F de \qquad (17)$$

where n_1 and n_2 are the numbers of molecules of each kind under consideration, and F is a function of the relative velocity and of the angle θ, and de depends on the limits of variation within which we class encounters as of the same kind.

Now let A describe the boundary of an element of volume dV while AB and $A'B'$ move parallel to themselves, then B, A', and B' will also describe equal and similar elements of volume.

The number of molecules of the first kind, the lines representing the velocities of which terminate in the element dV at A, will be

$$n_1 = f_1(a)dV \qquad (18)$$

The number of molecules of the second kind which have velocities corresponding to OB will be

$$n_2 = f_2(b)dV \qquad (19)$$

and the number of encounters of the given kind between these two sets of molecules will be

$$f_1(a)f_2(b)(dV)^2 F de \qquad (20)$$

The lines representing the velocities of these molecules after encounters of the given kind will terminate within elements of volume at A' and B', each equal to dV.

In like manner we should find for the number of encounters between molecules whose original velocities correspond to elements equal to dV described about A' and B', and whose subsequent velocities correspond to elements equal to dV described about A and B,

$$f_1(a')f_2(b')(dV)^2F'de \qquad (21)$$

where F' is the same function of $B'A'$ and $A'GA$ that F is of BA and AGA'. F is therefore equal to F'.

When the number of pairs of molecules which change their velocities from OA, OB to OA', OB' is equal to the number which change from OA', OB' to OA, OB, then the final distribution of velocity will be obtained, which will not be altered by subsequent exchanges. This will be the case when

$$f_1(a)f_2(b) = f_1(a')f_2(b') \qquad (22)$$

Now the only relation between a, b and a', b' is

$$M_1a^2 + M_2b^2 = M_1a'^2 + M_2b'^2 \qquad (23)$$

whence we obtain

$$f_1(a) = C_1e^{-(a^2/\alpha^2)}, \quad f_2(b) = C_2e^{-(b^2/\beta^2)} \qquad (24)$$

where

$$M_1\alpha^2 = M_2\beta^2 \qquad (25)$$

By integrating $\iiint C_1e^{-\frac{\xi^2+\eta^2+\zeta^2}{a^2}} d\xi d\eta d\zeta$, and equating the result to N_1, we obtain the value of C_1. If, therefore, the distribution of velocities among N_1 molecules is such that the number of molecules whose component velocities are between ξ and $\xi+d\xi$, η and $\eta+d\eta$, and ζ and $\zeta+d\zeta$ is

$$dN_1 = \frac{N_1}{\alpha^3\pi^{3/2}}e^{-\frac{\xi^2+\eta^2+\zeta^2}{\alpha^2}} d\xi d\eta d\zeta \qquad (26)$$

then this distribution of velocities will not be altered by the exchange of velocities among the molecules by their mutual action.

This is therefore a possible form of the final distribution of velocities. It is also the only form; for if there were any other, the exchange between velocities represented by OA and OA' would not be equal. Suppose that the number of molecules having velocity OA' increases at the expense of OA. Then since the total number of molecules corresponding to OA' remains constant, OA' must communicate as many to OA'', and so on till they return to OA.

Hence if OA, OA', OA'', etc. be a series of velocities, there will be a tendency of each molecule to assume the velocities OA, OA', OA'', etc. in order, returning to OA. Now it is impossible to assign a reason why the successive velocities of a molecule should be arranged in this cycle, rather than in the reverse order. If, therefore, the direct exchange between OA and OA' is not equal, the equality cannot be preserved by exchange in a cycle. Hence the direct exchange between OA and OA' is equal, and the distribution we have determined is the only one possible.

This final distribution of velocity is attained only when the molecules have had a great number of encounters, but the great rapidity with which the encounters succeed each other is such that in all motions and changes of the gaseous system except the most violent, the form of the distribution of velocity is only slightly changed.

When the gas moves in mass, the velocities now determined are compounded with the motion of translation of the gas.

When the differential elements of the gas are changing their figure, being compressed or extended along certain axes, the values of the mean square of the velocity will be different in different directions. It is probable that the form of the function will then be

$$f_1(\xi\eta\zeta) = \frac{N_1}{\alpha\beta\gamma\pi^{3/2}}e^{-\left(\frac{\xi^2}{\alpha^2}+\frac{\eta^2}{\beta^2}+\frac{\zeta^2}{\gamma^2}\right)} \qquad (27)$$

where α, β, γ are slightly different. I have not, however attempted to investigate the exact distribution of velocities in this case, as the theory of motion of gases does not require it.

When one gas is diffusing through another, or when heat is being conducted through a gas, the distribution of velocities will be different in the positive and negative directions, instead of being symmetrical, as in the case we have considered. The want of symmetry, however, may be treated as very small in most actual cases.

The principal conclusions which we may draw from this investigation are as follows. Calling α the modulus of velocity,

1st. The mean velocity is

$$\bar{v} = \frac{2}{\sqrt{\pi}}\,\alpha \tag{28}$$

2nd. The mean square of the velocity is

$$\overline{v^2} = \frac{3}{2}\alpha^2 \tag{29}$$

3rd. The mean value of ξ^2 is

$$\overline{\xi^2} = \frac{1}{2}\alpha^2 \tag{30}$$

4th. The mean value of ξ^4 is

$$\overline{\xi^4} = \frac{3}{4}\alpha^4 \tag{31}$$

5th. The mean value of $\xi^2\eta^2$ is

$$\overline{\xi^2\eta^2} = \frac{1}{4}\alpha^4 \tag{32}$$

6th. When there are two systems of molecules

$$M_1\alpha^2 = M_2\beta^2 \tag{33}$$

whence

$$M_1v_1^2 = M_2v_2^2 \tag{34}$$

or the mean *vis viva* of a molecule will be the same in each system. This is a very important result in the theory of gases, and it is independent of the nature of the action between the molecules, as are all the other results relating to the final distribution of velocities. We shall find that it leads to the law of gases known as that of Equivalent Volumes.

Variation of Functions of the Velocity due to encounters between the Molecules

We may now proceed to write down the values of $\delta\bar{Q}/\delta t$ in the different cases. We shall indicate the mean value of any quantity for all the molecules of one kind by placing a bar over the symbol which represents that quantity for any particular molecule, but in expressions where all such quantities are to be taken at their mean values, we shall, for convenience, omit the bar. We shall use the symbols δ_1 and δ_2 to indicate the effect produced by molecules of the first kind and second kind respectively, and δ_3 to indicate the effect of external forces. We shall also confine ourselves to the case in which $n = 5$, since it is not only free from mathematical difficulty, but is the only case which is consistent with the laws of viscosity of gases.

In this case V disappears, and we have for the effect of the second system on the first,

$$\frac{\delta Q}{\delta t} = N_2 \left\{ \frac{K(M_1+M_2)}{M_1 M_2} \right\}^{\frac{1}{4}} A \int_0^\pi (Q'-Q)d\phi \tag{35}$$

where the functions of ξ, η, ζ in $(Q'-Q)d\phi$ must be put equal to their mean values for all the molecules, and A_1 or A_2 must be put for A according as $\sin^2\theta$ or $\sin^2 2\theta$ occurs in the expressions in equations (4), (5), (6), (7). We thus obtain

$$(\alpha) \quad \frac{\delta_2\xi_1}{\delta t} = \left\{ \frac{K}{M_1 M_2(M_1+M_2)} \right\}^{\frac{1}{4}} N_2 M_2 A_1(\xi_2-\xi_1) \tag{36}$$

$$(\beta) \quad \frac{\delta_2 \xi_1^2}{\delta t} = \left\{ \frac{K}{M_1 M_2 (M_1 + M_2)} \right\}^{\frac{1}{2}} \frac{N_2 M_2}{M_1 + M_2}$$

$$\times \{ 2A_1 (\xi_2 - \xi_1)(M_1 \xi_1 + M_2 \xi_2)$$

$$+ A_2 M_2 (\overline{\eta_2 - \eta_1}^2 + \overline{\zeta_2 - \zeta_1}^2 - 2\overline{\xi_2 - \xi_1}^2) \} \qquad (37)$$

$$\frac{\delta_2 \xi_1 \eta_1}{\delta t} = \left\{ \frac{K}{M_1 M_2 (M_1 + M_2)} \right\}^{\frac{1}{2}} \frac{N_2 M_2}{M_1 + M_2}$$

$$[A_1 \{ 2M_2 \xi_2 \eta_2 - 2M_1 \xi_1 \eta_1$$

$$+ (M_1 - M_2)(\xi_1 \eta_2 + \xi_2 \eta_1) \} \qquad (38)$$

$$- 3A_2 M_2 (\xi_2 - \xi_1)(\eta_2 - \eta_1)]$$

$$(\gamma) \quad \frac{\delta_2 \xi_1 V_1^2}{\delta t} = \left\{ \frac{K}{M_1 M_2 (M_1 + M_2)} \right\}^{\frac{1}{2}} N_2 M_2 \left[A_1 \{ \overline{\xi_2 - \xi_1} V_1^2 \right.$$

$$+ 2\xi_1 (U - V_1^2) \}$$

$$+ \frac{M_2}{M_1 + M_2} (2A_1 - 3A_2) 2(\xi_2 - \xi_1) \times (U - V_1^2)$$

$$\qquad (39)$$

$$+ \frac{M_2}{M_1 + M_2} (2A_1 + 2A_2) \xi_1 V^2$$

$$+ \left(\frac{M_2}{M_1 + M_2} \right)^2 (2A_1 - 2A_2) \times 2(\xi_2 - \xi_1) V^2 \right]$$

using the symbol δ_2 to indicate variations arising from the action of molecules of the second system.

These are the values of the rate of variation of the mean values of ξ_1, ξ_1^2, $\xi_1 \eta_1$, and $\xi_1 V_1^2$, for the molecules of the first kind due to their encounters with molecules of the second kind. In all of them we must multiply up all functions of ξ, η, ζ, and take the mean values of the products so found. As this has to be done for all such functions, I have omitted the bar over each function in these expressions.

To find the rate of variation due to the encounters among the particles of the same system, we have only to alter the suffix $_{(2)}$ into $_{(1)}$ throughout, and to change K, the coefficient of the force between M_1 and M_2 into K_1, that of the force between two molecules of the first system. We thus find

$$(\alpha) \qquad \frac{\delta_1 \overline{\xi_1}}{\delta t} = 0 \qquad\qquad\qquad (40)$$

$$(\beta) \qquad \frac{\delta_1 \overline{\xi_1^2}}{dt} = \left(\frac{K_1}{2M_1^3}\right)^{\frac{1}{2}} M_1 N_1 A_2 \{\overline{\eta_1^2} + \overline{\zeta_1^2} - 2\overline{\xi_1^2} - (\overline{\eta_1 \cdot \eta_1}$$

$$+ \overline{\zeta_1 \cdot \zeta_1} - 2\overline{\xi_1 \cdot \xi_1})\} \qquad (41)$$

$$\frac{\delta_1 \overline{\xi_1 \eta_1}}{\delta t} = \left(\frac{K_1}{2M_1^3}\right)^{\frac{1}{2}} M_1 N_1 A_2 3\{\overline{\xi_1 \cdot \eta_1} - \overline{\xi_1 \eta_1}\} \qquad (42)$$

$$(\gamma) \qquad \frac{\delta_1 \overline{\xi_1 V_1^2}}{\delta t} = \left(\frac{K_1}{2M_1^3}\right) M_1 N_1 A_2 3(\overline{\xi_1 \cdot V_1^2} - \overline{\xi_1 V_1^2}) \qquad (43)$$

·   These quantities must be added to those in equations (36) to (39) in order to get the rate of variation in the molecules of the first kind due to their encounters with molecules of both systems. When there is only one kind of molecules, the latter equations give the rates of variation at once.

On the Action of External Forces on a System of Moving Molecules

We shall suppose the external force to be like the force of gravity, producing equal acceleration on all the molecules. Let the components of the force in the three coordinate directions be X, Y,

Z. Then we have by dynamics for the variations of ξ, ξ^2, and ξV^2 due to this cause,

$$(\alpha) \qquad \frac{\delta_3 \xi}{\delta t} = X \qquad\qquad (44)$$

$$(\beta) \qquad \frac{\delta_3 . \xi^2}{\delta t} = 2\xi X \qquad\qquad (45)$$

$$\frac{\delta_3 . \xi \eta}{\delta t} = \eta X + \xi Y \qquad\qquad (46)$$

$$(\gamma) \quad \frac{\delta_3 . \xi V^2}{\delta t} = 2\xi(\xi X + \eta Y + \zeta Z) + X V^2 \qquad (47)$$

where δ_3 refers to variations due to the action of external forces.

On the total rate of change of the different functions of the velocity of the molecules of the first system arising from their encounters with molecules of both systems and from the action of external forces

To find the total rate of change arising from these causes, we must add

$$\frac{\delta_1 Q}{\delta t}, \quad \frac{\delta_2 Q}{\delta t}, \quad \text{and} \quad \frac{\delta_3 Q}{\delta t}$$

the quantities already found. We shall find it, however, most convenient in the remainder of this investigation to introduce a change in the notation, and to substitute for

$$\xi, \eta, \text{and } \zeta, \quad u+\xi, \quad v+\eta, \text{ and } w+\zeta \qquad (48)$$

where u, v, and w are so chosen that they are the mean values of the components of the velocity of all molecules of the same system in the immediate neighbourhood of a given point. We shall also write

$$M_1 N_1 = \rho_1, \quad M_2 N_2 = \rho_2 \qquad (49)$$

c

where ρ_1 and ρ_2 are the densities of the two systems of molecules, that is, the mass in unit of volume. We shall also write

$$\left(\frac{K_1}{2M_1^2}\right)^{\frac{1}{2}} = k_1, \quad \left(\frac{K}{M_1M_2(M_1+M_2)}\right)^{\frac{1}{2}} = k, \quad \text{and} \quad \left(\frac{K_2}{2M_2^3}\right)^{\frac{1}{2}} = k_2 \tag{50}$$

ρ_1, ρ_2, k_1, k_2, and k are quantities the absolute values of which can be deduced from experiment. We have not as yet experimental data for determining M, N, or K.

We thus find for the rate of change of the various functions of velocity,

$$(\alpha) \quad \frac{\delta u_1}{\delta t} = kA_1\rho_2(u_2-u_1)+X \tag{51}$$

$$(\beta) \quad \frac{\delta.\xi_1^2}{\delta t} = k_1A_2\rho_1\{\eta_1^2+\zeta_1^2-2\xi_1^2\} + k\rho_2\frac{M_2}{M_1+M_2}$$
$$\times \{2A_1(u_2-u_1)^2$$
$$+A_2(\overline{v_2-v_1}^2+\overline{w_2-w_1}^2-\overline{2u_2-u_1}^2)\}$$
$$+\frac{k\rho_2}{M_1+M_2}\{2A_1(M_2\xi_2^2-M_1\xi_1^2)$$
$$+A_2M_2(\eta_1^2+\zeta_1^2-2\xi_1^2+\eta_2^2+\zeta_2^2-2\xi_2^2)\} \tag{52}$$

also

$$(\gamma) \quad \frac{\delta.\xi\eta}{\delta t} = -3k_1A_2\rho_1\xi_1\eta_1 + k\rho_2\frac{M_2}{M_1+M_2}$$
$$\times (2A_1-3A_2)(u_2-u_1)(v_2-v_1)$$
$$+\frac{k\rho_2}{M_1+M_2}\{2A_1(M_2\xi_2\eta_2-M_1\xi_1\eta_1)$$
$$-3A_2M_2(\xi_1\eta_1+\xi_2\eta_2)\} \tag{53}$$

(γ) As the expressions for the variation of functions of three dimensions in mixed media are complicated, and as we shall not have occasion to use them, I shall give the case of a single medium,

$$\frac{\delta}{\delta t}(\xi_1^3 + \xi_1\eta_1^2 + \xi_1\zeta_1^2) = -3k_1\rho_1A_2(\xi_1^3 + \xi_1\eta_1^2 + \xi_1\zeta_1^2)$$

$$+ X(3\xi_1^2 + \eta_1^2 + \zeta_1^2) + 2Y\xi_1\eta_1 + 2Z\xi_1\zeta_1 \quad (54)$$

Theory of a Medium composed of Moving Molecules

We shall suppose the position of every moving molecule referred to three rectangular axes, and that the component velocities of any one of them, resolved in the directions of x, y, z, are

$$u + \xi, \quad v + \eta, \quad w + \zeta$$

where u, v, w are the components of the mean velocity of all the molecules which are at a given instant in a given element of volume, and ξ, η, ζ are the components of the relative velocity of one of these molecules with respect to the mean velocity.

The quantities u, v, w may be treated as functions of x, y, z, and t, in which case differentiation will be expressed by the symbol d. The quantities ξ, η, ζ, being different for every molecule, must be regarded as functions of t for each molecule. Their variation with respect to t will be indicated by the symbol δ.

The mean values of ξ^2 and other functions of ξ, η, ζ for all the molecules in the element of volume may, however, be treated as functions of x, y, z, and t.

If we consider an element of volume which always moves with the velocities u, v, w, we shall find that it does not always consist of the same molecules, because molecules are continually passing through its boundary. We cannot therefore treat it as a mass moving with the velocity u, v, w, as is done in hydrodynamics, but we must consider separately the motion of each molecule. When we have occasion to consider the variation of the properties of this element during its motion as a function of the time we shall use the symbol ∂.

We shall call the velocities u, v, w the velocities of translation of the medium, and ξ, η, ζ the velocities of agitation of the molecules.

Let the number of molecules in the element $dx\,dy\,dz$ be $N\,dx\,dy\,dz$, then we may call N the number of molecules in unit of volume. If M is the mass of each molecule, and ρ the density of the element, then

$$MN = \rho \tag{55}$$

Transference of Quantities across a Plane Area

We must next consider the molecules which pass through a given plane of unit area in unit of time, and determine the quantity of matter, of momentum, of heat, etc. which is transferred from negative to the positive side of this plane in unit of time.

We shall first divide the N molecules in unit of volume into classes according to the value of ξ, η, and ζ for each, and we shall suppose that the number of molecules in unit of volume whose velocity in the direction of x lies between ξ and $\xi + d\xi$, η and $\eta + d\eta$, ζ and $\zeta + d\zeta$ is dN, dN will then be a function of the component velocities, the sum of which being taken for all the molecules will give N the total number of molecules. The most probable form of this function for a medium in its state of equilibrium is

$$dN = \frac{N}{\alpha^3 \pi^{\frac{3}{2}}} e^{-\frac{\xi^2 + \eta^2 + \zeta^2}{\alpha^2}} d\xi\,d\eta\,d\zeta \tag{56}$$

In the present investigation we do not require to know the form of this function.

Now let us consider a plane of unit area perpendicular to x moving with a velocity of which the part resolved parallel to x is u'. The velocity of the plane relative to the molecules we have been considering is $u' - (u + \xi)$, and since there are dN of these molecules in unit of volume it will overtake

$$\{u' - (u + \xi)\}dN$$

such molecules in unit of time, and the number of such molecules passing from the negative to the positive side of the plane, will be

$$(u + \xi - u')dN$$

Now let Q be any property belonging to the molecule, such as its mass, momentum, *vis viva*, etc., which it carries with it across the plane, Q being supposed a function of ξ or of ξ, η, and ζ, or to vary in any way from one molecule to another, provided it be the same for the selected molecules whose number is dN, then the quantity of Q transferred across the plane in the positive direction in unit of time is

$$\int(u - u' + \xi)QdN$$

or

$$(u - u')\int QdN + \int \xi QdN \qquad (57)$$

If we put $\bar{Q}N$ for $\int QdN$, and $\overline{\xi Q}N$ for $\int \xi QdN$, then we may call $\bar{Q}$ the mean value of Q, and $\overline{\xi Q}$ the mean value of ξQ, for all the particles in the element of volume, and we may write the expression for the quantity of Q which crosses the plane in unit of time

$$(u - u')\bar{Q}N + \overline{\xi Q}N \qquad (58)$$

(α) Transference of Matter across a Plane—Velocity of the Fluid

To determine the quantity of matter which crosses the plane, make Q equal to M the mass of each molecule; then, since M is the same for all molecules of the same kind, $\overline{M} = M$; and since the mean value of ξ is zero, the expression is reduced to

$$(u - u')MN = (u - u')\rho \qquad (59)$$

If $u = u'$, or if the plane moves with velocity u, the whole excess of matter transferred across the plane is zero; the velocity of the fluid may therefore be defined as the velocity whose components are u, v, w.

(β) Transference of Momentum across a Plane—System of Pressures at any point of the Fluid

The momentum of any one molecule in the direction of x is $M(u + \xi)$. Substituting this for Q, we get for the quantity of momentum transferred across the plane in the positive direction

$$(u - u')u\rho + \overline{\xi^2}\rho \tag{60}$$

If the plane moves with the velocity u, this expression is reduced to $\overline{\xi^2}\rho$, where $\overline{\xi^2}$ represents the mean value of ξ^2.

This is the whole momentum in the direction of x of the molecules projected from the negative to the positive side of the the plane in unit of time. The mechanical action between the parts of the medium on opposite sides of the plane consists partly of the momentum thus transferred, and partly of the direct attractions or repulsions between molecules on opposite sides of the plane. The latter part of the action must be very small in gases, so that we may consider the pressure between the parts of the medium on opposite sides of the plane as entirely due to the constant bombardment kept up between them. There will also be a transference of momentum in the directions of y and z across the same plane,

$$(u - u')v\rho + \overline{\xi\eta}\rho \tag{61}$$

and

$$(u - u')w\rho + \overline{\xi\zeta}\rho \tag{62}$$

where $\overline{\xi\eta}$ and $\overline{\xi\zeta}$ represent the mean values of these products.

If the plane moves with the mean velocity u of the fluid, the total force exerted on the medium on the positive side by the projection of molecules into it from the negative side will be

a normal pressure $\overline{\xi^2}\rho$ in the direction of x,

a tangential pressure $\overline{\xi\eta}\rho$ in the direction of y,

and a tangential pressure $\overline{\xi\zeta}\rho$ in the direction of z.

If X, Y, Z are the components of the pressure on unit of area of a plane whose direction cosines are l, m, n,

$$\left. \begin{array}{l} X = l\overline{\xi^2}\rho + m\overline{\xi\eta}\rho + n\overline{\xi\zeta}\rho \\ Y = l\overline{\xi\eta}\rho + m\overline{\eta^2}\rho + n\overline{\eta\zeta}\rho \\ Z = l\overline{\xi\zeta}\rho + m\overline{\eta\zeta}\rho + n\overline{\zeta^2}\rho \end{array} \right\} \tag{63}$$

When a gas is not in a state of violent motion the pressures in all directions are nearly equal, in which case, if we put

$$\overline{\xi^2}\rho + \overline{\eta^2}\rho + \overline{\zeta^2}\rho = 3p \tag{64}$$

the quantity p will represent the mean pressure at a given point, and $\overline{\xi^2}\rho$, $\overline{\eta^2}\rho$, and $\overline{\zeta^2}\rho$ will differ from p only by small quantities; $\overline{\eta\zeta}\rho$, $\overline{\zeta\xi}\rho$, and $\overline{\xi\eta}\rho$ will then be also small quantities with respect to p.

Energy in the Medium—Actual Heat

The actual energy of any molecule depends partly on the velocity of its centre of gravity, and partly on its rotation or other internal motion with respect to the centre of gravity. It may be written

$$\tfrac{1}{2}M\{(u+\xi)^2 + (v+\eta)^2 + (w+\zeta)^2\} + \tfrac{1}{2}EM \tag{65}$$

where $\tfrac{1}{2}EM$ is the internal part of the energy of the molecule, the form of which is at present unknown. Summing for all the molecules in unit of volume, the energy is

$$\tfrac{1}{2}(u^2+v^2+w^2)\rho + \tfrac{1}{2}(\overline{\xi^2}+\overline{\eta^2}+\overline{\zeta^2})\rho + \tfrac{1}{2}\bar{E}\rho \tag{66}$$

The first term gives the energy due to the motion of translation of the medium in mass, the second that due to the agitation of the centres of gravity of the molecules, and the third that due to the internal motion of the parts of each molecule.

If we assume with Clausius that the ratio of the mean energy of internal motion to that of agitation tends continually towards a definite value $(\beta - 1)$,† we may conclude that, except in very violent disturbances, this ratio is always preserved, so that

$$\bar{E} = (\beta - 1)(\xi^2 + \eta^2 + \zeta^2) \tag{67}$$

The total energy of the invisible agitation in unit of volume will then be

$$\tfrac{1}{2}\beta(\xi^2 + \eta^2 + \zeta^2) \tag{68}$$

or

$$\tfrac{3}{2}\beta p \tag{69}$$

This energy being in the form of invisible agitation, may be called the total heat in the unit of volume of the medium.

(γ) Transference of Energy across a Plane—Conduction of Heat

Putting

$$Q = \tfrac{1}{2}\beta(\xi^2 + \eta^2 + \zeta^2)M, \quad \text{and} \quad u = u' \tag{70}$$

we find for the quantity of heat carried over the unit of area by conduction in unit of time

$$\tfrac{1}{2}\beta(\overline{\xi^3} + \overline{\xi\eta^2} + \overline{\xi\zeta^2})\rho \tag{71}$$

where $\overline{\xi^3}$, etc. indicate the mean values of ξ^3, etc. They are always small quantities.

On the Rate of Variation of Q in an Element of Volume, Q being any property of the Molecules in that Element

Let Q be the value of the quantity for any particular molecule, and $\bar{Q}$ the mean value of Q for all the molecules of the same kind within the element.

The quantity $\bar{Q}$ may vary from two causes. The molecules within the element may by their mutual action or by the action of external forces produce an alteration of $\bar{Q}$, or molecules may pass

---

† [Vol. 1, Selection 8.]

into the element and out of it, and so cause an increase or diminution of the value of $\bar{Q}$ within it. If we employ the symbol δ to denote the variation of Q due to actions of the first kind on the individual molecules, and the symbol ∂ to denote the actual variation of Q in an element moving with the mean velocity of the system of molecules under consideration, then by the ordinary investigation of the increase or diminution of matter in an element of volume as contained in treatises on Hydrodynamics,

$$\frac{\partial \bar{Q}N}{\partial t} = \frac{\delta \bar{Q}}{\delta t}N - \frac{d}{dx}\{(u-u')\bar{Q}N + \overline{\xi Q}N\}$$
$$- \frac{d}{dy}\{(v-v')\bar{Q}N + \overline{\eta Q N}\} - \frac{d}{dz}\{(w-w')\bar{Q}N + \overline{\zeta Q N}\} \quad (72)$$

where the last three terms are derived from equation (58) and two similar equations, and denote the quantity of Q which flows out of an element of volume, that element moving with the velocities u', v', w'. If we perform the differentiations and then make $u' = u$, $v' = v$, and $w' = w$, then the variation will be that in an element which moves with the actual mean velocity of the system of molecules, and the equation becomes

$$\frac{\partial \bar{Q}N}{\delta t} + \bar{Q}N\left(\frac{du}{dx}+\frac{dv}{dy}+\frac{dw}{dz}\right) + \frac{d}{dx}(\overline{\xi Q}N) + \frac{d}{dy}(\overline{\eta Q}N)$$
$$+ \frac{d}{dz}(\overline{\zeta Q}N) = \frac{\delta Q}{\delta t}N \quad (73)$$

Equation of Continuity

Put $Q = M$ the mass of a molecule; M is unalterable, and we have, putting $MN = \rho$,

$$\frac{\partial \rho}{\partial t}+\rho\left(\frac{du}{dx}+\frac{dv}{dy}+\frac{dw}{dz}\right) = 0 \quad (74)$$

C*

which is the ordinary equation of continuity in hydrodynamics, the element being supposed to move with the velocity of the fluid. Combining this equation with that from which it was obtained, we find

$$N\frac{\partial\bar{Q}}{\partial t}+\frac{d}{dx}(\overline{\xi Q}N)+\frac{d}{dy}(\overline{\eta Q}N)+\frac{d}{dz}(\overline{\zeta Q}N) = N\frac{\delta Q}{\delta t} \tag{75}$$

a more convenient form of the general equation.

Equations of Motion (α)

To obtain the Equation of Motion in the direction of x, put $Q = M_1(u_1+\xi_1)$, the momentum of a molecule in the direction of x.

We obtain the value of $\delta Q/\delta t$ from equation (51), and the equation may be written

$$\left.\begin{aligned}\rho_1\frac{\partial u_1}{\partial t}+\frac{d}{dx}(\rho_1\overline{\xi_1^2})+\frac{d}{dy}(\rho_1\overline{\xi_1\eta_1})+\frac{d}{dz}(\rho_1\overline{\xi_1\zeta_1})\\ = kA_1\rho_1\rho_2(u_2-u_1)+X\rho_1\end{aligned}\right\} \tag{76}$$

In this equation the first term denotes the efficient force per unit of volume, the second the variation of normal pressure, the third and fourth the variations of tangential pressure, the fifth the resistance due to the molecules of a different system, and the sixth the external force acting on the system.

The investigation of the values of the second, third, and fourth terms must be deferred till we consider the variations of the second degree.

Condition of Equilibrium of a Mixture of Gases

In a state of equilibrium u_1 and u_2 vanish, $\rho_1\xi_1^2$ becomes p_1, and the tangential pressures vanish, so that the equation becomes

$$\frac{dp_1}{dx} = X\rho_1 \tag{77}$$

which is the equation of equilibrium in ordinary hydrostatics.

This equation, being true of the system of molecules forming the first medium independently of the presence of the molecules of the second system, shews that if several kinds of molecules are mixed together, placed in a vessel and acted on by gravity, the final distribution of the molecules of each kind will be the same as if none of the other kinds had been present. This is the same mode of distribution as that which Dalton considered to exist in a mixed atmosphere in equilibrium, the law of diminution of density of each constituent gas being the same as if no other gases were present.†

This result, however, can only take place after the gases have been left for a considerable time perfectly undisturbed. If currents arise so as to mix the strata, the composition of the gas will be made more uniform throughout.

The result at which we have arrived as to the final distribution of gases, when left to themselves, is independent of the law of force between the molecules.

Diffusion of Gases

If the motion of the gases is slow, we may still neglect the tangential pressures. The equation then becomes for the first system of molecules

$$\rho_1 \frac{\partial u_1}{\partial t} + \frac{dp_1}{dx} = kA_1\rho_1\rho_2(u_2 - u_1) + X\rho_1 \tag{78}$$

and for the second,

$$\rho_2 \frac{\partial u_2}{\partial t} + \frac{dp_2}{dx} = kA_1\rho_1\rho_2(u_1 - u_2) + X\rho_2 \tag{79}$$

In all cases of quiet diffusion we may neglect the first term of each equation. If we then put $p_1 + p_2 = p$, and $\rho_1 + \rho_2 = \rho$, we find by adding,

$$\frac{dp}{dx} = X\rho \tag{80}$$

---

† [J. Dalton, *Memoirs of the Manchester Philosophical Society* **5**, 535 (1802).]

If we also put $p_1u_1 + p_2u_2 = pu$, then the volumes transferred in opposite directions across a plane moving with velocity u will be equal, so that

$$p_1(u_1 - u) = p_2(u - u_2) = \frac{p_1p_2}{pp_1\rho_2kA_1} \cdot \left(X\rho_1 - \frac{dp_1}{dx}\right) \qquad (81)$$

Here $p_1\,(u_1 - u)$ is the volume of the first gas transferred in unit of time across unit of area of the plane reduced to pressure unity, and at the actual temperature; and $p_2\,(u - u_2)$ is the equal volume of the second gas transferred across the same area in the opposite direction.

The external force X has very little effect on the quiet diffusion of gases in vessels of moderate size. We may therefore leave it out in our definition of the coefficient of diffusion of two gases.

When two gases not acted on by gravity are placed in different parts of a vessel at equal pressures and temperatures, there will be mechanical equilibrium from the first, and u will always be zero. This will also be approximately true of heavy gases, provided the denser gas is placed below the lighter. Mr Graham has described in his paper on the Mobility of Gases,† experiments which were made under these conditions. A vertical tube had its lower tenth part filled with a heavy gas, and the remaining nine-tenths with a lighter gas. After the lapse of a known time the upper tenth part of the tube was shut off, and the gas in it analyzed, so as to determine the quantity of the heavier gas which had ascended into the upper tenth of the tube during the given time.

In this case we have

$$u = 0 \qquad (82)$$

$$p_1u_1 = -\frac{p_1p_2}{\rho_1\rho_2kA_1}\frac{1}{p}\frac{dp_1}{dx} \qquad (83)$$

and by the equation of continuity,

$$\frac{dp_1}{dt} + \frac{d}{dx}(p_1u_1) = 0 \qquad (84)$$

---

† *Philosophical Transactions*, 1863 [p. 385].

whence

$$\frac{dp_1}{dt} = \frac{p_1 p_2}{\rho_1 \rho_2 k A_1} \frac{1}{p} \frac{d^2 p_1}{dx^2} \tag{85}$$

or if we put $D = \frac{p_1 p_2}{\rho_1 \rho_2 k A_1} \frac{1}{p}$,

$$\frac{dp_1}{dt} = D \frac{d^2 p_1}{dx^2} \tag{86}$$

The solution of this equation is

$$p_1 = C_1 + C_2 e^{-n^2 Dt} \cos(nx + \alpha) + \&c. \tag{87}$$

If the length of the tube is a, and if it is closed at both ends,

$$p_1 = C_1 + C_2 e^{-\frac{\pi^2 D}{a^2} t} \cos \frac{\pi x}{a} + C_3 e^{-4 \frac{\pi^2 D}{a^2} t} \cos 2\frac{\pi x}{a} + \&c. \tag{88}$$

where C_1, C_2, C_3 are to be determined by the condition that when $t = 0$, $p_1 = p$, from $x = 0$ to $x = \frac{1}{10}a$ and $p_1 = 0$ from $x = \frac{1}{10}a$ to $x = a$. The general expression for the case in which the first gas originally extends from $x = 0$ to $x = b$, and in which after a time t the gas from $x = 0$ to $x = c$ is collected, is

$$\frac{p_1}{p} = \frac{b}{a} + \frac{2a}{\pi^2 c} \left\{ e^{-\frac{\pi^2 D}{a^2} t} \sin \frac{\pi b}{a} \sin \frac{\pi c}{a} \right.$$
$$\left. + \frac{1}{2^2} e^{-4\frac{\pi^2 D}{a^2} t} \sin \frac{2\pi b}{a} \sin \frac{2\pi c}{a} + \&c. \right\} \tag{89}$$

where p_1/p is the proportion of the first gas to the whole in the portion from $x = 0$ to $x = c$.

In Mr Graham's experiments, in which one-tenth of the tube was filled with the first gas, and the proportion of the first gas in the tenth of the tube at the other end ascertained after a time t, this proportion will be

$$\frac{p_1}{p} = \frac{1}{10} - \frac{20}{\pi^2} \left\{ e^{-\frac{\pi^2 D}{a^2} t} \sin^2 \frac{\pi}{10} - e^{-2^2 \frac{\pi^2 D}{a^2} t} \sin^2 2\frac{\pi}{10} \right.$$
$$\left. + e^{-3^2 \frac{\pi^2 D}{a^2} t} \sin^2 3\frac{\pi}{10} - \&c. \right\} \tag{90}$$

We find for a series of values of p_1/p taken at equal intervals of time T, where

$$T = \frac{\log_e 10}{10\pi^2} \frac{a^2}{D}$$

Time	$\dfrac{p_1}{p}$
0	0
T	·01193
$2T$	·02305
$3T$	·03376
$4T$	·04366
$5T$	·05267
$6T$	·06072
$8T$	·07321
$10T$	·08227
$12T$	·08845
∞	·10000

Mr Graham's experiments on carbonic acid and air, when compared with this Table give $T = 500$ seconds nearly for a tube 0·57 metre long. Now

$$D = \frac{\log_e 10}{10\pi^2} \frac{a^2}{T} \qquad (91)$$

whence

$$D_i = ·0235$$

for carbonic acid and air, in inch-grain-second measure.

Definition of the Coefficient of Diffusion

D is the volume of gas reduced to unit of pressure which passes in unit of time through unit of area when the total pressure is uniform and equal to p, and the pressure of either gas increases or diminishes by unity in unit of distance. D may be called the coefficient of diffusion. It varies directly as the square of the absolute temperature, and inversely as the total pressure p.

The dimensions of D are evidently $L^2 T^{-1}$, where L and T are the standards of length and time.

In considering this experiment of the interdiffusion of carbonic acid and air, we have assumed that air is a simple gas. Now it is well known that the constituents of air can be separated by mechanical means, such as passing them through a porous diaphragm, as in Mr Graham's experiments on Atmolysis. The discussion of the interdiffusion of three or more gases leads to a much more complicated equation than that which we have found for two gases, and it is not easy to deduce the coefficients of interdiffusion of the separate gases. It is therefore to be desired that experiments should be made on the interdiffusion of every pair of the more important pure gases which do not act chemically on each other, the temperature and pressure of the mixture being noted at the time of experiment.

Mr Graham has also published in Brande's *Journal* for 1829, pt. 2, p. 74,† the results of experiments on the diffusion of various gases out of a vessel through a tube into air. The coefficients of diffusion deduced from these experiments are:

Air and Hydrogen	·026216
Air and Marsh-gas	·010240
Air and Ammonia	·00962
Air and Olefiant gas	·00771
Air and Carbonic acid	·00682
Air and Sulphurous acid	·00582
Air and Chlorine	·00486

The value for carbonic acid is only one-third of that deduced from the experiment with the vertical column. The inequality of composition of the mixed gas in different parts of the vessel is, however, neglected; and the diameter of the tube at the middle part, where it was bent, was probably less than that given.

---

† [*Quarterly Journal of Science.*]

Those experiments on diffusion which lasted ten hours, all give smaller values of D than those which lasted four hours, and this would also result from the mixture of the gases in the vessel being imperfect.

Interdiffusion through a small hole

When two vessels containing different gases are connected by a small hole, the mixture of gases in each vessel will be nearly uniform except near the hole; and the inequality of the pressure of each gas will extend to a distance from the hole depending on diameter of the hole, and nearly proportional to that diameter.

Hence in the equation

$$\rho_1 \frac{\partial u_1}{\partial t} + \frac{dp_1}{dx} = kA\rho_1\rho_2(u_2 - u_1) + X\rho \tag{92}$$

the term dp_1/dx will vary inversely as the diameter of the hole, while u_1 and u_2 will not vary considerably with the diameter.

Hence when the hole is very small the right-hand side of the equation may be neglected, and the flow of either gas through the hole will be independent of the flow of the other gas, as the term $kA\rho_1p_2(u_2 - u_1)$ becomes comparatively insignificant.

One gas therefore will escape through a very fine hole into another nearly as fast as into a vacuum; and if the pressures are equal on both sides, the volumes diffused will be as the square roots of the specific gravities inversely, which is the law of diffusion of gases established by Graham.†

Variation of the invisible agitation (β)

By putting for Q in equation (75)

$$Q = \frac{M}{2}\{(u_1 + \xi_1)^2 + (v_1 + \eta_1)^2 + (w_1 + \zeta_1)^2 + (\beta - 1)(\xi_1^2 + \eta_1^2 + \zeta_1^2)\}$$

$$\tag{93}$$

---

† *Trans. Royal Society of Edinburgh*, Vol. XII, p. 22 [(1834)].

and eliminating by means of equations (76) and (52), we find

$$
\begin{aligned}
&\tfrac{1}{2}\rho_1\frac{\partial}{\partial t}\beta_1(\xi_1^2+\eta_1^2+\zeta_1^2)+\rho_1\xi_1^2\frac{du_1}{dx}+\rho_1\eta_1^2\frac{dv_1}{dy}+\rho_1\zeta_1^2\frac{dw_1}{dz}\\[2mm]
&\quad+\rho_1\eta_1\zeta_1\left(\frac{dv_1}{dz}+\frac{dw_1}{dy}\right)\\[2mm]
&\quad+\rho_1\zeta_1\xi_1\left(\frac{dw_1}{dx}+\frac{du_1}{dz}\right)+\rho_1\xi_1\eta_1\left(\frac{du_1}{dy}+\frac{dv_1}{dx}\right)\\[2mm]
&\quad+\beta_1\left\{\frac{d}{dx}(\rho_1\xi_1^3+\rho_1\xi_1\eta_1^2+\rho_1\xi_1\zeta_1^2)\right.\\[2mm]
&\quad+\frac{d}{dy}(\rho_1\eta_1\xi_1^2+\rho_1\eta_1^3+\rho_1\eta_1\zeta_1^2)\\[2mm]
&\quad+\frac{d}{dz}(\rho_1\zeta_1\xi_1^2+\rho_1\zeta_1\eta_1^2+\rho_1\zeta_1^3)\Big\}\\[2mm]
&\quad=\frac{k\rho_1\rho_2 A_1}{M_1+M_2}[M_2\{(u_2-u_1)^2+(v_2-v_1)^2+(w_2-w_1)^2\}\\[2mm]
&\quad+M_2(\xi_2^2+\eta_2^2+\zeta_2^2)-M_1(\xi_1^2+\eta_1^2+\zeta_1^2)]
\end{aligned}
\tag{94}
$$

In this equation the first term represents the variation of invisible agitation or heat; the second, third, and fourth represent the cooling by expansion; the fifth, sixth, and seventh the heating effect of fluid friction or viscosity; and the last the loss of heat by conduction. The quantities on the other side of the equation represent the thermal effects of diffusion, and the communication of heat from one gas to the other.

The equation may be simplified in various cases, which we shall take in order.

1st. Equilibrium of Temperature between two Gases—
Law of Equivalent Volumes

We shall suppose that there is no motion of translation, and no transfer of heat by conduction through either gas. The equation (94) is then reduced to the following form,

$$\tfrac{1}{2}\rho_1 \frac{\partial}{\partial t}\beta_1(\xi_1^2+\eta_1^2+\zeta_1^2) = \frac{k\rho_1\rho_2 A_1}{M_1+M_2}\{M_2(\xi_2^2+\eta_2^2+\zeta_2^2) \\ -M_1(\xi_1^2+\eta_1^2+\zeta_1^2)\} \qquad (95)$$

If we put

$$\frac{M_1}{M_1+M_2}(\xi_1^2+\eta_1^2+\zeta_1^2) = Q_1, \quad \text{and} \quad \frac{M_2}{M_1+M_2}(\xi_2^2+\eta_2^2+\zeta_2^2) = Q_2 \qquad (96)$$

we find

$$\frac{\partial}{\partial t}(Q_2-Q_1) = -\frac{2kA_1}{M_1+M_2}(M_2\rho_2\beta_1+M_1\rho_1\beta_2)(Q_2-Q_1) \qquad (97)$$

or

$$Q_2-Q_1 = Ce^{-nt}, \text{ where } n = \frac{2kA_1}{M_1+M_2}(M_2\rho_2\beta_2+M_1\rho_1\beta_1)\frac{1}{\beta_1\beta_2} \qquad (98)$$

If, therefore, the gases are in contact and undisturbed, Q_1 and Q_2 will rapidly become equal. Now the state into which two bodies come by exchange of invisible agitation is called equilibrium of heat or equality of temperature. Hence when two gases are at the same temperature,

$$Q_1 = Q_2 \qquad (99)$$

or

$$1 = \frac{Q_1}{Q_2} = \frac{M_1(\xi_1^2+\eta_1^2+\zeta_1^2)}{M_2(\xi_2^2+\eta_2^2+\zeta_2^2)}$$

$$= \frac{M_1\dfrac{p_1}{\rho_1}}{M_2\dfrac{p_2}{\rho_2}}$$

Hence if the pressures as well as the temperatures be the same in two gases,

$$\frac{M_1}{\rho_1} = \frac{M_2}{\rho_2} \tag{100}$$

or the masses of the individual molecules are proportional to the density of the gas.

This result, by which the relative masses of the molecules can be deduced from the relative densities of the gases, was first arrived at by Gay-Lussac from chemical considerations. It is here shewn to be a necessary result of the Dynamical Theory of Gases; and it is so, whatever theory we adopt as to the nature of the action between the individual molecules, as may be seen by equation (34), which is deduced from perfectly general assumptions as to the nature of the law of force.

We may therefore henceforth put s_1/s_2 for M_1/M_2, where s_1, s_2 are the specific gravities of the gases referred to a standard gas.

If we use θ to denote the temperature reckoned from absolute zero of a gas thermometer, M_0 the mass of a molecule of hydrogen, V_0^2 its mean square of velocity at temperature unity, s the specific gravity of any other gas referred to hydrogen, then the mass of a molecule of the other gas is

$$M = M_0 s \tag{101}$$

Its mean square of velocity,

$$V^2 = \frac{1}{s} V_0^2 \theta \tag{102}$$

Pressure of the gas,

$$p = \tfrac{1}{3}\frac{\rho}{s}\theta V_0^2 \tag{103}$$

We may next determine the amount of cooling by expansion.

Cooling by Expansion

Let the expansion be equal in all directions, then

$$\frac{du}{dx} = \frac{dv}{dy} = \frac{dw}{dz} = -\frac{1}{3\rho}\frac{\partial \rho}{\partial t} \tag{104}$$

and du/dy and all terms of unsymmetrical form will be zero.

If the mass of gas is of the same temperature throughout there will be no conduction of heat, and the equation (94) will become

$$\tfrac{1}{2}\rho\beta\frac{\partial \bar{V}^2}{\partial t} - \tfrac{1}{3}\bar{V}^2\frac{\partial \rho}{\partial t} = 0 \tag{105}$$

or

$$2\frac{\partial \rho}{\rho} = 3\beta\frac{\partial \bar{V}^2}{\bar{V}^2} = 3\beta\frac{\partial \theta}{\theta} \tag{106}$$

or

$$\frac{\partial \theta}{\theta} = \frac{2}{3\beta}\frac{\partial \rho}{\rho} \tag{107}$$

which gives the relation between the density and the temperature in a gas expanding without exchange of heat with other bodies. We also find

$$\frac{\partial p}{p} = \frac{\partial \rho}{\rho} + \frac{\partial \theta}{\theta}$$

$$= \frac{2+3\beta}{3\beta}\frac{\partial \rho}{\rho} \tag{108}$$

which gives the relation between the pressure and the density.

Specific Heat of Unit of Mass at Constant Volume

The total energy of agitation of unit of mass is $\tfrac{1}{2}\beta V^3 = \tfrac{1}{2}E$, or

$$E = \frac{3\beta}{2}\frac{p}{\rho} \tag{109}$$

If, now, additional energy in the form of heat be communicated to it without changing its density,

$$\partial E = \frac{3\beta}{2}\frac{\partial p}{\rho} = \frac{3\beta}{2}\frac{p}{\rho}\frac{\partial\theta}{\theta} \tag{110}$$

Hence the specific heat of unit of mass at constant volume is in dynamical measure

$$\frac{\partial E}{\partial\theta} = \frac{3\beta}{2}\frac{p}{\rho\theta} \tag{111}$$

Specific Heat of Unit of Mass at Constant Pressure

By the addition of the heat ∂E the temperature was raised $\partial\theta$ and the pressure ∂p. Now, let the gas expand without communication of heat till the pressure sinks to its former value, and let the final temperature be $\theta + \partial'\theta$. The temperature will thus sink by a quantity $\partial\theta - \partial'\theta$, such that

$$\frac{\partial\theta - \partial'\theta}{\theta} = \frac{2}{2+3\beta}\frac{\partial p}{p} = \frac{2}{2+3\beta}\frac{\partial\theta}{\theta}$$

whence

$$\frac{\partial'\theta}{\theta} = \frac{3\beta}{2+3\beta}\frac{\partial\theta}{\theta} \tag{112}$$

and the specific heat of unit of mass at constant pressure is

$$\frac{\partial E}{\partial'\theta} = \frac{2+3\beta}{2}\frac{p}{\rho\theta} \tag{113}$$

The ratio of the specific heat at constant pressure to that of constant volume is known in several cases from experiment. We shall denote this ratio by

$$\gamma = \frac{2+3\beta}{3\beta} \tag{114}$$

whence

$$\beta = \tfrac{2}{3}\frac{1}{\gamma-1} \tag{115}$$

The specific heat of unit of volume in ordinary measure is at constant volume

$$\frac{1}{\gamma-1}\frac{p}{J\theta} \tag{116}$$

and at constant pressure

$$\frac{\gamma}{\gamma-1}\frac{p}{J\theta} \tag{117}$$

where J is the mechanical equivalent of unit of heat.

From these expressions Dr Rankine† has calculated the specific heat of air, and has found the result to agree with the value afterwards determined experimentally by M. Regnault‡.

Thermal Effects of Diffusion

If two gases are diffusing into one another, then, omitting the terms relating to heat generated by friction and to conduction of heat, the equation (94) gives

$$\left.\begin{aligned}
\tfrac{1}{2}\rho_1\frac{\partial}{\partial t}&\beta_1(\xi_1^2+\eta_1^2+\zeta_1^2)+\tfrac{1}{2}\rho_2\frac{\partial}{\partial t}\beta_2(\xi_2^2+\eta_2^2+\zeta_2^2)\\
&+p_1\left(\frac{du_1}{dx}+\frac{dv_1}{dy}+\frac{dw_1}{dz}\right)+p_2\left(\frac{du_2}{dx}+\frac{dv_2}{dy}+\frac{dw_2}{dz}\right)\\
&=k\rho_1\rho_2A_1\{(u_1-u_2)^2+(v_1-v_2)^2+(w_1-w_2)^2\}
\end{aligned}\right\} \tag{118}$$

By comparison with equations (78) and (79), the right-hand side of this equation becomes

$$X(\rho_1u_1+\rho_2u_2)+Y(\rho_1v_1+\rho_2v_2)+Z(\rho_1w_1+\rho_2w_2)$$

$$-\left(\frac{dp_1}{dx}u_1+\frac{dp_1}{dy}v_1+\frac{dp_1}{dz}w_1\right)-\left(\frac{dp_2}{dx}u_2+\frac{dp_2}{dy}v_2+\frac{dp_2}{dz}w_2\right)$$

$$-\tfrac{1}{2}\rho_1\frac{\partial}{\partial t}(u_1^2+v_1^2+w_1^2)-\tfrac{1}{2}\rho_2\frac{\partial}{\partial t}(u_2^2+v_2^2+w_2^2)$$

---

† *Transactions of the Royal Society of Edinburgh*. Vol. XX (1850) [p. 147].
‡ *Comptes Rendus* [Academie des Sciences, Paris], 1853 [**36**, 676].

The equation (118) may now be written

$$
\left.
\begin{aligned}
&\tfrac{1}{2}\rho_1 \frac{\partial}{\partial t}\{u_1^2+v_1^2+w_1^2+\beta_1(\xi_1^2+\eta_1^2+\zeta_1^2)\} \\[2mm]
&\quad +\tfrac{1}{2}\rho_2 \frac{\partial}{\partial t}\{u_2^2+v_2^2+w_2^2+\beta_2(\xi_2^2+\eta_2^2+\zeta_2^2)\} \\[2mm]
&= X(\rho_1 u_1+\rho_2 u_2)+Y(\rho_1 v_1+\rho_2 v_2) \\[2mm]
&\quad +Z(\rho_1 w_1+\rho_2 w_2)-\left(\frac{d.pu}{dx}+\frac{d.pv}{dy}+\frac{d.pw}{dz}\right)
\end{aligned}
\right\}
\tag{119}
$$

The whole increase of energy is therefore that due to the action of the external forces *minus* the cooling due to the expansion of the mixed gases. If the diffusion takes place without alteration of the volume of the mixture, the heat due to the mutual action of the gases in diffusion will be exactly neutralized by the cooling of each gas as it expands in passing from places where it is dense to places where it is rare.

Determination of the Inequality of Pressure in different Directions due to the Motion of the Medium

Let us put

$$
\rho_1\xi_1^2 = p_1+q_1 \quad \text{and} \quad \rho_2\xi_2^2 = p_2+q_2
\tag{120}
$$

Then by equation (52),

$$
\left.
\begin{aligned}
\frac{\delta q_1}{\delta t} &= -3k_1 A_2\rho_1 q_1-\frac{k}{M_1+M_2}(2M_1 A_1+3M_2 A_2)\rho_2 q_1 \\[2mm]
&\quad -k(3A_2-2A_1)\frac{M_1}{M_1+M_2}\rho_1 q_2-k\rho_1\rho_2 \\[2mm]
&\quad \times \frac{M_2}{M_1+M_2}(A_2-\tfrac{2}{3}A_1)(2\overline{u_1-u_2}^2-\overline{v_1-v_2}^2-\overline{w_1-w_2}^2)
\end{aligned}
\right\}
\tag{121}
$$

the last term depending on diffusion; and if we omit in equation (75) terms of three dimensions in ξ, η, ζ, which relate to conduction of heat, and neglect quantities of the form $\xi\eta\rho$ and $\rho\xi^2-p$, when not multiplied by the large coefficients k, k_1, and k_2, we get

$$\frac{\partial q}{\partial t} + 2p\frac{du}{dx} - \tfrac{2}{3}p\left(\frac{du}{dx} + \frac{dv}{dy} + \frac{dw}{dz}\right) = \frac{\delta q}{\delta t} \tag{122}$$

If the motion is not subject to any very rapid changes, as in all cases except that of the propagation of sound, we may neglect $\partial q/\partial t$. In a single system of molecules

$$\frac{\delta q}{\delta t} = -3kA_2\rho q \tag{123}$$

whence

$$q = -\frac{2p}{3kA_2\rho}\left\{\frac{du}{dx} - \tfrac{1}{3}\left(\frac{du}{dx} + \frac{dv}{dy} + \frac{dw}{dz}\right)\right\} \tag{124}$$

If we make

$$\tfrac{1}{3}\frac{1}{kA_2}\frac{p}{\rho} = \mu \tag{125}$$

μ will be the coefficient of viscosity, and we shall have by equation (120),

$$\left.\begin{aligned}
\rho\xi^2 &= p - 2\mu\left\{\frac{du}{dx} - \tfrac{1}{3}\left(\frac{du}{dx} + \frac{dv}{dy} + \frac{dw}{dz}\right)\right\} \\[2mm]
\rho\eta^2 &= p - 2\mu\left\{\frac{dv}{dy} - \tfrac{1}{3}\left(\frac{du}{dx} + \frac{dv}{dy} + \frac{dw}{dz}\right)\right\} \\[2mm]
\rho\zeta^2 &= p - 2\mu\left\{\frac{dw}{dz} - \tfrac{1}{3}\left(\frac{du}{dx} + \frac{dv}{dy} + \frac{dw}{dz}\right)\right\}
\end{aligned}\right\} \tag{126}$$

and by transformation of co-ordinates we obtain

$$\left.\begin{aligned}
\rho\eta\zeta &= -\mu\left(\frac{dv}{dz} + \frac{dw}{dy}\right) \\[2mm]
\rho\zeta\xi &= -\mu\left(\frac{dw}{dx} + \frac{du}{dz}\right) \\[2mm]
\rho\xi\eta &= -\mu\left(\frac{du}{dy} + \frac{dv}{dx}\right)
\end{aligned}\right\} \tag{127}$$

These are the values of the normal and tangential stresses in a simple gas when the variation of motion is not very rapid, and when μ, the coefficient of viscosity, is so small that its square may be neglected.

Equations of Motion corrected for Viscosity

Substituting these values in the equation of motion (76), we find

$$\rho \frac{\partial u}{\partial t} + \frac{dp}{dx} - \mu \left\{ \frac{d^2u}{dx^2} + \frac{d^2u}{dy^2} + \frac{d^2u}{dz^2} \right\} - \tfrac{1}{3}\mu \frac{d}{dx} \left(\frac{du}{dx} + \frac{dv}{dy} + \frac{dw}{dz} \right) = X\rho \tag{128}$$

with two other equations which may be written down with symmetry. The form of these equations is identical with that of those deduced by Poisson[†] from the theory of elasticity, by supposing the strain to be continually relaxed at a rate proportional to its amount. The ratio of the third and fourth terms agrees with that given by Professor Stokes.[‡]

If we suppose the inequality of pressure which we have denoted by q to exist in the medium at any instant, and not to be maintained by the motion of the medium, we find, from equation (123),

$$q_1 = Ce^{-3kA_2\rho t} \tag{129}$$

$$= Ce^{-t/T} \quad \text{if} \quad T = \frac{1}{3kA_2\rho} = \frac{\mu}{p} \tag{130}$$

the stress q is therefore relaxed at a rate proportional to itself, so that

$$\frac{\delta q}{q} = \frac{\delta t}{T} \tag{131}$$

We may call T the modulus of the time of relaxation.

[†] *Journal de l'École Polytechnique*, 1829, Tom. XIII, Cah. XX, p. 139.

[‡] " On the Friction of Fluids in Motion and the Equilibrium and Motion of Elastic Solids ", *Cambridge Phil. Trans.* Vol. VIII (1845) p. 297, equation (2).

If we next make $k = 3$, so that the stress q does not become relaxed, the medium will be an elastic solid, and the equation

$$\frac{\partial(\rho \xi^2 - p)}{\partial t} + 2p\frac{du}{dx} - \tfrac{2}{3}p\left(\frac{du}{dx} + \frac{dv}{dy} + \frac{dw}{dz}\right) = 0 \qquad (132)$$

may be written

$$\frac{\partial}{\partial t}\left\{(p_{xx} - p) + 2p\frac{d\alpha}{dx} - \tfrac{2}{3}p\left(\frac{d\alpha}{dx} + \frac{d\beta}{dy} + \frac{d\gamma}{dz}\right)\right\} = 0 \qquad (133)$$

where α, β, γ are the displacements of an element of the medium, and p_{xx} is the normal pressure in the direction of x. If we suppose the initial value of this quantity zero, and p_{xx} originally equal to p, then, after a small displacement,

$$p_{xx} = p - p\left(\frac{d\alpha}{dx} + \frac{d\beta}{dy} + \frac{d\gamma}{dz}\right) - 2p\frac{d\alpha}{dx} \qquad (134)$$

and by transformation of co-ordinates the tangential pressure

$$p_{xy} = -p\left(\frac{d\alpha}{dy} + \frac{d\beta}{dx}\right) \qquad (135)$$

The medium has now the mechanical properties of an elastic solid, the rigidity of which is p, while the cubical elasticity is $\tfrac{5}{3}p$.†

The same result and the same ratio of the elasticities would be obtained if we supposed the molecules to be at rest, and to act on one another with forces depending on the distance, as in the statical molecular theory of elasticity. The coincidence of the properties of a medium in which the molecules are held in equilibrium by attractions and repulsions, and those of a medium in which the molecules move in straight lines without acting on each other at all, deserve notice from those who speculate on theories of physics.

---

† *Camb. Phil. Trans.* Vol. VIII (1845) p. 311, equation (29).

The fluidity of our medium is therefore due to the mutual action of the molecules, causing them to be deflected from their paths.

The coefficient of instantaneous rigidity of a gas is therefore p

The modulus of the time of relaxation is T

The coefficient of viscosity is $\mu = pT$

$$(136)$$

Now p varies as the density and temperature conjointly, while T varies inversely as the density.

Hence μ varies as the absolute temperature, and is independent of the density.

This result is confirmed by the experiments of Mr Graham on the Transpiration of Gases,† and by my own experiments on the Viscosity or Internal Friction of Air and other Gases.‡

The result that the viscosity is independent of the density, follows from the Dynamical Theory of Gases, whatever be the law of force between the molecules. It was deduced by myself§ from the hypothesis of hard elastic molecules, and M. O. E. Meyer‖ has given a more complete investigation on the same hypothesis.

The experimental result, that the viscosity is proportional to the absolute temperature, requires us to abandon this hypothesis, which would make it vary as the square root of the absolute temperature, and to adopt the hypothesis of a repulsive force inversely as the fifth power of the distance between the molecules, which is the only law of force which gives the observed result.

Using the foot, the grain, and the second as units, my experiments gave for the temperature of 62° Fahrenheit, and in dry air,

$$\mu = 0.0936$$

---

† *Philosophical Transactions*, 1846 [p. 573] and 1849 [p. 349].

‡ *Proceedings of the Royal Society*, February 8, 1866 [**15**, 14]; *Philosophical Transactions*, 1866, p. 249.

§ *Philosophical Magazine*, January 1860 [Vol. 1, Selection 10].

‖ Poggendorff's *Annalen*, 1865 [*Ann. Phys.* **125**, 177, 401, 564].

If the pressure is 30 inches of mercury, we find, using the same units,

$$p = 477360000$$

Since $pT = \mu$, we find that the modulus of the time of relaxation of rigidity in air of this pressure and temperature is

$$\frac{1}{5099100000} \text{ of a second}$$

This time is exceedingly small, even when compared with the period of vibration of the most acute audible sounds; so that even in the theory of sound we may consider the motion as steady during this very short time, and use the equations we have already found, as has been done by Professor Stokes.†

Viscosity of a Mixture of Gases

In a complete mixture of gases, in which there is no diffusion going on, the velocity at any point is the same for all the gases.

Putting

$$\tfrac{2}{3}\left(2\frac{du}{dx} - \frac{dv}{dy} - \frac{dw}{dz}\right) = U \tag{137}$$

equation (122) becomes

$$\left.\begin{aligned}
p_1 U = -3k_1 A_2 \rho_1 q_1 &- \frac{k}{M_1 + M_2}(2M_1 A_1 + 3M_2 A_2)\rho_2 q_1 \\
&- k(3A_2 - 2A_1)\frac{M_2}{M_1 + M_2}\rho_1 q_2
\end{aligned}\right\} \tag{138}$$

Similarly,

$$\left.\begin{aligned}
p_2 U = -3k_2 A_2 \rho_2 q_2 &- \frac{k}{M_1 + M_2}(2M_2 A_1 + 3M_1 A_2)\rho_1 q_2 \\
&- k(3A_2 - 2A_1)\frac{M_1}{M_1 + M_2}\rho_2 q_1
\end{aligned}\right\} \tag{139}$$

---

† " On the effect of the internal Friction of Fluids on the motion of Pendulums ", *Cambridge* [*Philosophical*] *Transactions*, Vol. IX (1850), art. 79.

Since $p = p_1 + p_2$ and $q = q_1 + q_2$, where p and q refer to the mixture, we shall have

$$\mu U = -q = -(q_1 + q_2)$$

where μ is the coefficient of viscosity of the mixture.

If we put s_1 and s_2 for the specific gravities of the two gases, referred to a standard gas, in which the values of p and q at temperature θ_0 and p_0 and ρ_0,

$$\mu = \frac{p_0 \theta}{\rho_0 \theta_0} \cdot \frac{E p_1^2 + F p_1 p_2 + G p_2^2}{3A_2 k_1 s_1 E p_1^2 + H p_1 p_2 + 3A_2 k_2 s_2 G p_2^2} \qquad (140)$$

where μ is the coefficient of viscosity of the mixture, and

$$\left.\begin{array}{l} E = \dfrac{k s_1}{s_1 + s_2}(2s_2 A_1 + 3s_1 A_2) \\[2mm] F = 3A_2(k_1 s_1 + k_2 s_2) - (3A_2 - 2A_1)k\dfrac{2s_1 s_2}{s_1 + s_2} \\[2mm] G = \dfrac{k s_2}{s_1 + s_2}(2s_1 A_1 + 3s_2 A_2) \\[2mm] H = 3A_2 s_1 s_2(3k_1 k_2 A_2 + 2k^2 A_1) \end{array}\right\} \qquad (141)$$

This expression is reduced to μ_1 when $p_2 = 0$, and to μ_2 when $p_1 = 0$. For other values of p_1 and p_2 we require to know the values of k, the coefficient of mutual interference of the molecules of the two gases. This might be deduced from the observed values of μ for mixtures, but a better method is by making experiments on the interdiffusion of the two gases. The experiments of Graham on the transpiration of gases, combined with my experiments on the viscosity of air, give as values of k_1 for air, hydrogen, and carbonic acid,

Air $\qquad k_1 = \quad 4\cdot81 \times 10^{10}$,

Hydrogen $\qquad k_1 = 142\cdot8 \ \times 10^{10}$,

Carbonic acid $k_1 = \quad 3\cdot9 \ \times 10^{10}$.

The experiments of Graham in 1863,† on the interdiffusion of air and carbonic acid, give the coefficient of mutual interference of these gases,

$$\text{Air and carbonic acid } k = 5{\cdot}2 \times 10^{10};$$

and by taking this as the absolute value of k, and assuming that the ratios of the coefficients of interdiffusion given at page 67 are correct, we find

$$\text{Air and hydrogen } k = 29{\cdot}8 \times 10^{10}.$$

These numbers are to be regarded as doubtful, as we have supposed air to be a simple gas in our calculations, and we do not know the value of k between oxygen and nitrogen. It is also doubtful whether our method of calculation applies to experiments such as the earlier observations of Mr Graham.

I have also examined the transpiration-times determined by Graham for mixtures of hydrogen and carbonic acid, and hydrogen and air, assuming a value of k roughly, to satisfy the experimental results about the middle of the scale. It will be seen that the calculated numbers for hydrogen and carbonic acid exhibit the peculiarity observed in the experiments, that a small addition of hydrogen *increases* the transpiration-time of carbonic acid, and that in both series the times of mixtures depend more on the slower than on the quicker gas.

The assumed values of k in these calculations were:

$$\text{For hydrogen and carbonic acid } k = 12{\cdot}5 \times 10^{10},$$

$$\text{For hydrogen and air} \qquad\quad k = 18{\cdot}8 \times 10^{10}$$

and the results of observation and calculation are, for the times of transpiration of mixtures, given in the table.

The numbers given are the ratios of the transpiration-times of mixtures to that of oxygen as determined by Mr Graham, compared with those given by the equation (140) deduced from our theory.

---

† *Philosophical Transactions*, 1863 [p. 385].

Hydrogen and Carbonic acid		Observed	Calculated	Hydrogen and Air		Observed	Calculated
100	0	·4321	·4375	100	0	·4434	·4375
97·5	2·5	·4714	·4750	95	5	·5282	·5300
95	5	·5157	·5089	90	10	·5880	·6028
90	10	·5722	·5678	75	25	·7488	·7438
75	25	·6786	·6822	50	50	·8179	·8488
50	50	·7339	·7652	25	75	·8790	·8946
25	75	·7535	·7468	10	90	·8880	·8983
10	90	·7521	·7361	5	95	·8960	·8996
0	100	·7470	·7272	0	100	·9000	·9010

Conduction of Heat in a Single Medium (γ)

The rate of conduction depends on the value of the quantity

$$\tfrac{1}{2}\beta\rho(\xi^3 + \xi\eta^2 + \xi\zeta^2)$$

where ξ^3, $\xi\eta^2$, and $\xi\zeta^2$ denote the mean values of those functions of ξ, η, ζ for all the molecules in a given element of volume.

As the expressions for the variations of this quantity are somewhat complicated in a mixture of media, and as the experimental investigation of the conduction of heat in gases is attended with great difficulty, I shall confine myself here to the discussion of a single medium.

Putting

$$Q = M(u+\xi)\{u^2+v^2+w^2+2u\xi+2v\eta+2w\zeta+\beta(\xi^2+\eta^2+\zeta^2)\} \tag{142}$$

and neglecting terms of the forms $\xi\eta$ and ξ^3 and $\xi\eta^2$ when not multiplied by the large coefficient k_1, we find by equations (75), (77), and (54),

$$\left. \begin{aligned} &\rho\frac{\partial}{\partial t}\beta(\xi^3 + \xi\eta^2 + \xi\zeta^2) + \beta\frac{d}{dx}.\rho(\xi^4 + \xi^2\eta^2 + \xi^2\zeta^2) \\ &-\beta(\xi^2+\eta^2+\zeta^2)\frac{dp}{dx} - 2\beta\xi^2\frac{dp}{dx} \\ &= -3k_1\rho^2 A_2\beta\{\xi^3 + \xi\eta^2 + \xi\zeta^2\} \end{aligned} \right\} \tag{143}$$

The first term of this equation may be neglected, as the rate of conduction will rapidly establish itself. The second term contains quantities of four dimensions in ξ, η, ζ, whose values will depend on the distribution of velocity among the molecules. If the distribution of velocity is that which we have proved to exist when the system has no external force acting on it and has arrived at its final state, we shall have by equations (29), (31), (32),

$$\overline{\xi^4} = \overline{3\xi^2} . \overline{\xi^2} = 3\frac{p^2}{\rho^2} \tag{144}$$

$$\overline{\xi^2\eta^2} = \overline{\xi^2} . \overline{\eta^2} = \frac{p^2}{\rho^2} \tag{145}$$

$$\overline{\xi^2\zeta^2} = \overline{\xi^2} . \overline{\zeta^2} = \frac{p^2}{\rho^2} \tag{146}$$

and the equation of conduction may be written

$$5\beta\frac{p^2}{\rho\theta}\frac{d\theta}{dx} = -3k_1\rho^2 A_2\beta\{\xi^3 + \xi\eta^2 + \xi\zeta^2\} \tag{147}$$

[Addition made December 17, 1866.]

[Final Equilibrium of Temperature]

[The left-hand side of equation (147), as sent to the Royal Society, contained a term $2(\beta-1)p/\rho \, dp/dx$, the result of which was to indicate that a column of air, when left to itself, would assume a temperature varying with the height, and greater above than below. The mistake arose from an error† in equation (143). Equation (147), as now corrected, shews that the flow of heat depends on the variation of temperature only, and not on the direction of the variation of pressure. A vertical column would therefore, when in thermal equilibrium have the same temperature throughout.

When I first attempted this investigation I overlooked the fact that $\overline{\xi^4}$ is not the same as $\overline{\xi^2} . \overline{\xi^2}$, and so obtained as a result that the temperature diminishes as the height increases at a greater rate than it does by expansion when air is carried up in mass. This

---

† The last term on the left-hand side was not multiplied by β.

leads at once to a condition of instability, which is inconsistent with the second law of thermodynamics. I wrote to Professor Sir W. Thomson about this result, and the difficulty I had met with, but presently discovered *one* of my mistakes, and arrived at the conclusion that the temperature would increase with the height. This does not lead to mechanical instability, or to any self-acting currents of air, and I was in some degree satisfied with it. But it is equally inconsistent with the second law of thermodynamics. In fact, if the temperature of any substance, when in thermic equilibrium, is a function of the height, that of any other substance must be the same function of the height. For if not, let equal columns of the two substances be enclosed in cylinders impermeable to heat, and put in thermal communication at the bottom. If, when in thermal equilibrium, the tops of the two columns are at different temperatures, an engine might be worked by taking heat from the hotter and giving it up to the cooler, and the refuse heat would circulate round the system till it was all converted into mechanical energy, which is in contradiction to the second law of thermodynamics.

The result as now given is, that temperature in gases, when in thermal equilibrium, is independent of height, and it follows from what has been said that temperature is independent of height in all other substances.

If we accept this law of temperature as the actual one, and examine our assumptions, we shall find that unless $\overline{\xi^4} = 3\overline{\xi^2} \cdot \overline{\xi^2}$, we should have obtained a different result. Now this equation is derived from the law of distribution of velocities to which we were led by independent considerations. We may therefore regard this law of temperature, if true, as in some measure a confirmation of the law of distribution of velocities.]

Coefficient of Conductivity

If C is the coefficient of conductivity of the gas for heat, then the quantity of heat which passes through unit of area in unit of time measured as mechanical energy, is

D

$$C\frac{d\theta}{dx} = \tfrac{5}{6}\frac{\beta}{k_1 A_2}\frac{p^2}{\rho^2 \theta}\frac{d\theta}{dx} \tag{148}$$

by equation (147).

Substituting for β its value in terms of γ by equation (115), and for k_1 its value in terms of μ by equation (125), and calling p_0, ρ_0, and θ_0 the simultaneous pressure, density, and temperature of the standard gas, and s the specific gravity of the gas in question, we find†

$$C = \frac{5}{3(\gamma-1)}\frac{p_0}{\rho_0 \theta_0}\frac{\mu}{s} \tag{149}$$

For air we have $\gamma = 1\cdot409$, and at the temperature of melting ice, or $274°\cdot6$ C. above absolute zero, $\sqrt{(p/\rho)} = 918\cdot6$ feet per second, and at $16°\cdot6$ C., $\mu = 0\cdot0936$ in foot-grain-second measure. Hence for air at $16°\cdot6$ C. the conductivity for heat is

$$C = 1172 \tag{150}$$

That is to say, a horizontal stratum of air one foot thick, of which the upper surface is kept at $17°$ C., and the lower at $16°$ C., would in one second transmit through every square foot of horizontal surface a quantity of heat the mechanical energy of which is equal to that of 2344 grains moving at the rate of one foot per second.

Principal Forbes‡ has deduced from his experiments on the conduction of heat in bars, that a plate of wrought iron one foot thick, with its opposite surfaces kept $1°$ C. different in temperature, would, when the mean temperature is $25°$ C., transmit in one minute through every square foot of surface as much heat as would raise one cubic foot of water $0\cdot0127°$C.

Now the dynamical equivalent in foot-grain-second measure of

---

† [According to Boltzmann, Selection 2, p. 141, the numerical constant should be 5/2 rather than 5/3. See also H. Poincaré, *Compt. rend. Acad. Sci.*, Paris, **116**, 1020 (1893).]

‡ " Experimental Inquiry into the Laws of the Conduction of Heat in Bars ", *Edinburgh* [*Royal Society*] *Transactions*, 1861–2 [**23**, 133].

the heat required to raise a cubic foot of water 1° C. is 1.9157×10^{10}.

It appears from this that iron at 25° C. conducts heat 3525 times better than air at 16°·6 C.

M. Clausius, from a different form of the theory, and from a different value of μ, found that lead should conduct heat 1400 times better than air. Now iron is twice as good a conductor of heat as lead, so that this estimate is not far different from that of M. Clausius in actual value.

In reducing the value of the conductivity from one kind of measure to another, we must remember that its dimensions are MLT^{-3}, when expressed in absolute dynamical measure.

Since all the quantities which enter into the expression for C are constant except μ, the conductivity is subject to the same laws as the viscosity, that is, it is independent of the pressure, and varies directly as the absolute temperature. The conductivity of iron diminishes as the temperature increases.

Also, since γ is nearly the same for air, oxygen, hydrogen and carbonic oxide, the conductivity of these gases will vary as the ratio of the viscosity to the specific gravity. Oxygen, nitrogen, carbonic oxide, and air will have equal conductivity, while that of hydrogen will be about seven times as great.

The value of γ for carbonic acid is 1·27, its specific gravity is $\frac{11}{8}$ of oxygen, and its viscosity $\frac{8}{11}$ of that of oxygen. The conductivity of carbonic acid for heat is therefore about $\frac{7}{9}$ of that of oxygen or of air.

2

Further Studies on the Thermal Equilibrium of Gas Molecules *

LUDWIG BOLTZMANN

SUMMARY

According to the mechanical theory of heat, the thermal properties of gases and other substances obey perfectly definite laws in spite of the fact that these substances are composed of large numbers of molecules in states of rapid irregular motion. The explanation of these properties must be based on probability theory, and for this purpose it is necessary to know the distribution function which determines the number of molecules in each state at every time. In order to determine this distribution function, $f(x, t) =$ number of molecules having energy x at time t, a partial differential equation for f is derived by considering how it changes during a small time interval as a result of collisions among molecules. If there are no external forces, and conditions are uniform throughout the gas, this equation takes the form (equation (16)):

$$\frac{\partial f(x, t)}{\partial t} = \int_0^\infty \int_0^{x+x'} \left[\frac{f(\xi, t)}{\sqrt{\xi}} \frac{f(x+x'-\xi, t)}{\sqrt{(x+x'-\xi)}} - \frac{f(x, t)}{\sqrt{x}} \frac{f(x't)}{\sqrt{x'}} \right]$$
$$\sqrt{(xx')} \; \psi(x, x', \xi) \; dx' \, d\xi$$

where the variables x and x' denote the energies of two molecules before a collision, and ξ and $(x+x'-\xi)$ denote their energies after the collision; $\psi(x, x', \xi)$ is a function which depends on the nature of the forces between the molecules.

If the velocity distribution is given by Maxwell's formula

$$f(x, t) = \text{(constant)} \; \sqrt{(x)} \; e^{-hx}$$

---

* [Originally published under the title " Weitere Studien über das Wärmegleichgewicht unter Gasmolekülen ", in *Sitzungsberichte Akad. Wiss.*, Vienna, part II, **66**, 275–370 (1872); reprinted in Boltzmann's *Wissenschaftliche Abhandlungen*, Vol. I, Leipzig, J. A. Barth, 1909, pp. 316–402.]

then the expression in brackets in the above equation will vanish, and the time-derivative of $f(x, t)$ will be zero. This is essentially the result already obtained in another way by Maxwell: once this velocity distribution has been reached, it will not be disturbed by collisions.

With the aid of the partial differential equation for f, we are able to go further and prove that if the distribution of states is not Maxwellian, it will tend toward the Maxwellian distribution as time goes on. This proof consists in showing that a quantity defined in terms of f,

$$E = \int\limits_{0}^{\infty} f(x, t) \left[\log \left(\frac{f(x, t)}{x} \right) - 1 \right] dx$$

can never increase but must always decrease or remain constant, if f satisfies the above differential equation. [This statement is now known as the H-theorem.] E must approach a minimum value and remain constant thereafter, and the corresponding final value of f will be the Maxwell distribution. Since E is closely related to the thermodynamic entropy in the final equilibrium state, our result is equivalent to a proof that the entropy must always *increase* or remain constant, and thus provides a microscopic interpretation of the second law of thermodynamics.

To clarify the reasoning involved in these proofs, we replace the continuous energy variable x by a discrete variable which can take only the values ϵ, 2ϵ, $3\epsilon, \ldots$; we then show that the same result can be derived by taking the limit $\epsilon \to 0$.

The differential equation for f is also given for the case when the velocity distribution may vary from one place to another, and all directions of velocity are not equivalent (equation (44)). For the special case of intermolecular forces varying inversely as the fifth power of the distance, this equation has a simple exact solution, and the coefficients of viscosity, heat conduction, and diffusion may be calculated. The results are essentially the same as those found by Maxwell [Selection 1].

The above results are generalized to gases composed of polyatomic molecules.

The mechanical theory of heat assumes that the molecules of a gas are not at rest, but rather are in the liveliest motion. Hence, even though the body does not change its state, its individual molecules are always changing their states of motion, and the various molecules take up many different positions with respect to each other. The fact that we nevertheless observe completely definite laws of behaviour of warm bodies is to be attributed to the circumstance that the most random events, when they occur in the same proportions, give the same average value. For the molecules of the body are indeed so numerous, and their motion is so rapid, that we can perceive nothing more than average values. One might compare the regularity of these average values with the amazing constancy of the average numbers provided by statistics, which are also derived from processes each of which is determined by a completely unpredictable interaction with many other factors. The molecules are likewise just so many individuals having the most varied states of motion, and it is only because the number of them that have, on the average, a particular state of motion is constant, that the properties of the gas remain unchanged. The determination of average values is the task of probability theory. Hence, the problems of the mechanical theory of heat are also problems of probability theory. It would, however, be erroneous to believe that the mechanical theory of heat is therefore afflicted with some uncertainty because the principles of probability theory are used. One must not confuse an incompletely known law, whose validity is therefore in doubt, with a completely known law of the calculus of probabilities; the latter, like the result of any other calculus, is a necessary consequence of definite premises, and is confirmed, insofar as these are correct, by experiment, provided sufficiently many observations have been made, which is always the case in the mechanical theory of heat because of the enormous number of molecules involved. It is only doubly imperative to handle the conclusions with the greatest strictness. If one does not merely wish to guess a few occasional values of the quantities that occur in gas theory, but rather desires to work with an exact theory, then he must first of all determine the probabilities of the various states which a given molecule will have during a very long time, or which different molecules will have at the same time. In other words, one must find the number of molecules out of the total number whose states lie between any given limits. Maxwell and I have previously treated this problem in several papers, without so far managing to obtain a complete solution of it. In fact, the problem seems to be very difficult in the case where each molecule consists of several mass-points (atoms), since one cannot integrate the equations of motion even for a complex of three atoms. Yet on closer consideration, it appears not unlikely that these probabilities can be obtained from the equations of motion alone, without having to integrate them. For the many simple laws of gases show that the expression for such probabilities must have certain general properties independent of the special nature of the gas, and such general properties can frequently be deduced from the equations of motion, without the necessity of integrating them. In fact, I have succeeded in finding the solution for gas molecules consisting of an arbitrary number of atoms. However, for the sake of a better over-all view of the subject, I will first treat the simplest case, where each molecule is a single point mass. I will then treat the general case, for which the calculation is similar.

I. Consideration of Monatomic Gas Molecules

Let a space be filled with many gas molecules, each of which is a simple point mass. Each molecule moves most of the time in a straight line with uniform velocity. Only when two molecules happen to come very close do they begin to interact with each other. I call this process, during which two molecules interact with each other, a collision of the two molecules, without implying a collision of elastic bodies; the force that acts during the collision may be completely arbitrary. Even when all the molecules initially have the same velocity, they will not retain the same velocity during the course of time. As a result of collisions, many molecules will acquire larger velocities and others will come to have smaller velocities, until finally a distribution of velocities among the molecules is established such that it is not changed by further collisions. In this final distribution, in general all possible velocities from zero up to a very large velocity will occur. The number of molecules whose velocity lies between r and $r + dr$ we shall call $F(r)dr$. The function F determines the velocity distribution completely. For the case of monatomic molecules, which we are now considering, Maxwell already found the value $Av^2 e^{-Bv^2}$ for $F(r)$, where A and B are constants, so that the probability of different velocities is given by a formula similar to that for the probability of different errors of observation in the theory of the method of least squares. The first proof which Maxwell gave for this formula was recognized to be incorrect even by himself. He later gave a very elegant proof that, if the above distribution has once been established, it will not be changed by collisions. He also tries to prove that it is the only velocity distribution that has this property. But the latter proof appears to me to contain a false inference.† It has still not yet been proved that, whatever the initial state of the gas may be, it must always approach the limit found by Maxwell. It is possible that there may be other possible limits. This proof is easily obtained, however, by the method which I am about to explain, and which also has the advantage that it permits one to deal directly with polyatomic molecules and thus with the case that probably occurs in nature.

I begin by defining the problem precisely again. Suppose therefore that we have a space R in which are found many gas molecules. Each molecule is a simple point mass, which moves in the

---

† First, Maxwell should really have proved that as many pairs of molecules will change their velocities from OA, OB to OA', OB' as conversely, while he actually only discusses whether a molecule changes its velocity from OA to OA' as often as it changes from OA' to OA. He then asserts that if the velocity OA were to change to OA' more often than the converse, then OA' would have to change to OA'' more often than the converse, by the same amount, since otherwise the number of molecules with velocity OA' could not remain constant. In fact, however, one can only conclude that one or more velocities OA'', OA''', etc., exist into which OA' is transformed more often than the converse. In order to prove finally that it is not possible for the velocity of a molecule to change from OA to OA' more often than the converse, Maxwell says, that otherwise there would be a recurrent cycle of velocities OA, OA', OA'' . . . OA which would be more often traversed in one direction than the opposite direction. But this cannot be, since there is no reason why a molecule should prefer to go around the cycle in one direction rather than the other. This latter assertion seems to me, however, to be one that should be proved rather than taken as already established. For if we take it as already proved that the velocity of a molecule changes from OA to OA' as often as the converse, then of course there is no reason why this cycle is more likely to be traversed in one direction than the other. If we assume, on the contrary, that the theorem to be proved is not yet known to be true, then the fact that the velocity of a molecule is more likely to change from OA to OA' than the converse, and more likely to change from OA' to OA'' than the converse, and so forth, would provide a reason why the cycle is more likely to be traversed in one direction than the other. The two processes are no less than identical. One cannot conclude therefore that they are *a priori* equally probable.

way already described. During the largest part of the time, it moves in a straight line with uniform velocity. Two molecules interact only when they come very close together. The law of the force that acts duting collisions must of course be given. However, I will not make any restrictive assumptions about this force law. Perhaps the two molecules rebound from each other like elastic spheres; perhaps any other force law may be given. As for the wall of the container that encloses the gas, I will assume that it reflects the molecules like elastic spheres. Any arbitrary force law would lead to the same formulae. However, it simplifies the matter if we make this special assumption about the container. We now set ourselves the following problem: suppose that initially ($t = 0$) the position, velocity, and velocity direction of each molecule is given. What is the position, velocity, and velocity direction of each molecule after an arbitrary time t has elapsed? Since the form of the container R as well as the force law for the collisions is given, this problem is of course completely determined. It is clear, however, that it is not completely soluble in this degree of generality. The solution would be much easier to find if, instead of this general problem, we set ourselves a rather more special one. We take account of only two conditions that pertain to the nature of the subject. First, it is clear that after a very long time each direction for the velocity of a molecule is equally likely. If it is only a question of finding the velocity distribution that will be established after a long time, then we can assume that already at the beginning each velocity direction is equally probable. In the most general case one still arrives at the same final state-distribution as in this special case. This is the first condition that we shall make. The second one is that the velocity distribution should already be uniform initially. I must next clarify what I mean by a uniform velocity distribution. For the following it will be better to use the kinetic energy of a molecule rather than its velocity. We shall now do this. Let x be the kinetic energy of our gas molecule, so that $x = mv^2/2$. R is the total space in which our gas is enclosed. We construct inside this space R a smaller space called r, whose shape is completely arbitrary but whose volume will be

D•

equal to one. We assume that in the space r there are a large number of molecules, so that its dimensions are large compared to the average distance of two neighbouring molecules; this imposes no real limitation, since we may choose the unit of volume to be as large as we like. The number of molecules in the space r whose kinetic energy at time t lies between x and $x + dx$ I will call $f(x, t)dx$. This number will in general depend on where I have chosen to construct r in R. For example, one might find the fast molecules on the right side and the slower ones on the left side of R. Then the number $f(x, t)dx$ would be different, depending on whether the space r was on the right or the left of R. When this is not the case, but rather $f(x, t)dx$ is the same at a given time no matter where r is, then I say that the distribution of kinetic energy is uniform. In other words, molecules with different kinetic energies are uniformly mixed with each other. The faster ones are not on the right, nor the slower ones on the left, nor conversely. It is again clear that after a very long time the distribution of kinetic energy will become uniform, for then each position in the gas is equivalent. The presence of the walls does not disturb this uniformity, for the molecules are reflected at them like elastic spheres; they come back from the wall as if the space on the other side of the wall were filled with another gas having the same properties. Hence, we may assume that the velocity distribution is already uniform initially. This assumption, and the equal probability of all directions of velocity initially, are the two restrictive conditions under which we shall treat the problem. It is clear that these two conditions will be satisfied for all following times, so that the state of the gas at time t will be completely determined by the function $f(x, t)$. If the state of the gas is given to us at an initial time, i.e. $f(x, 0)$, then we must find the state after an arbitrary time t has elapsed, namely $f(x, t)$. The way that we shall proceed is the same way that one always proceeds in such cases. We first calculate how much the function $f(x, t)$ changes during a very small time τ; from this we obtain a partial differential equation for the function $f(x, t)$; this must then be integrated in such a way that for $t = 0$, f takes the given value $f(x, 0)$. We have therefore a double task before us, first the establishment of the partial differential equation, and second its integration. We now turn to the first problem. $f(x, t)dx$ is the number of molecules in unit volume whose kinetic energy at time t lies between x and $x+dx$. As long as a molecule collides with no other molecules, it retains the same kinetic energy. If there were no collisions, $f(x, t)$ would not change; this function changes only because of collisions. If we wish to find the change in this function during a very short time τ, then we must consider the collisions during this time. We consider a collision, before which the kinetic energy of one of the colliding molecules is

between x and $x+dx$

and that of the other is

between x' and $x'+dx'$

Of course the nature of the collision is by no means completely determined yet. Depending on whether the collision is head-on or more or less glancing, the kinetic energy of one of the colliding molecules can have many different values after the collision. If we assume that this kinetic energy after the collision lies

between ξ and $\xi+d\xi$

then the kinetic energy of the second molecule after the collision is determined. If we denote the latter by ξ', then according to the principle of conservation of kinetic energy,

$$x+x' = \xi+\xi' \tag{1}$$

so that the sum of the kinetic energies of the two molecules is the same before and after the collision. We can now represent the limits between which the variables characterizing our collision lie by the following scheme:

	a	b	
Before the collision	$x, x+dx$	$x', x'+dx'$	(A)
After the collision	$\xi, \xi+d\xi$		

The kinetic energy of one molecule is in the column labelled a, and that of the other in the column labelled b. We now ask, how many collisions take place in time τ in unit volume such that the kinetic energies of the colliding molecules lie between the limits (A)? We shall denote this number by dn. The determination of dn can only be accomplished properly by considering the relative velocity of the two molecules. Since this calculation, although tedious, is not at all difficult, and has no special interest, and the result is so simple that one might almost say it is obvious, I will simply state the result. It is the following: dn is, first, proportional to the time τ; the longer this time τ is, the more collisions of the specified kind will occur, as long as τ is very small, so that the state of the gas during τ does not change noticeably. Second, dn is proportional to the quantity $f(x, t)dx$; this is indeed the number of molecules in unit volume whose kinetic energy lies between x and $x + dx$; the more such molecules there are in unit volume, the more often will they collide. Third, dn is proportional to $f(x', t)dx'$; for whatever holds for one of the colliding molecules will of course hold for the other one. The product of these three quantities must be multiplied by a certain proportionality factor, which is easily seen to be an infinitesimal like $d\xi$. This factor will in general depend on the nature of the collision, and hence on x, x' and ξ. To express these properties we shall write the proportionality factor as $d\xi \cdot \psi(x, x', \xi)$ so that we have therefore

$$dn = f(x, t)dx . f(x', t)dx' \, d\xi . \psi(x, x', \xi) \qquad (2)$$

This is the result obtained from an exact treatment of the collision process; this treatment also gives the function ψ, of course, as soon as the force law is given, since ψ depends on the force law. Since we do not need to know this function ψ yet, it would be superfluous to determine it here. We shall now keep x constant in the expression (2) for dn, and integrate x' and ξ over all possible values of those quantities, i.e. ξ from zero to $x + x'$ and x' from zero to infinity. The result of this integration will be called $\int dn$; thus

$$\int dn = \tau f(x, t)dx \int_0^\infty \int_0^{x+x'} f(x', t)\psi(x, x', \xi)dx'd\xi$$

Since x is to be considered constant for both integrations, we can write $f(x, t)$ inside the integral signs and obtain:†

$$\int dn = \tau dx \int_0^\infty \int_0^{x+x'} f(x, t)f(x', t)\psi(x, x', \xi)dx'd\xi \qquad (3)$$

What is this quantity $\int dn$? We have kept x constant. The kinetic energy of one molecule before the collision is still between

---

† Instead of actually writing out the limits of a definite integral, one can determine them in various ways, for example through inequalities. In the definite integral of equation (3) x is to be considered constant. The two integration variables are x' and ξ; these can take only positive values, including zero, for they represent kinetic energy; and indeed we must require $x+x'-\xi\geqq0$, since $x+x'-\xi$ is the kinetic energy of the second molecule after the collision. On the other hand, it is clear that all positive x' and ξ for which $x+x'-\xi$ also comes out to be positive represent possible collisions, and therefore lie within the limits of integration. The three inequalities

$$x\geqq0, \; \xi\geqq0, \; x+x'-\xi\geqq0 \qquad (3a)$$

define therefore the limits of integration of the integral in equation (3). This method of determining the limits is recommended by the fact that it often significantly shortens the calculation. One transforms the variables of integration to rectangular coordinate axes and determines the surface over

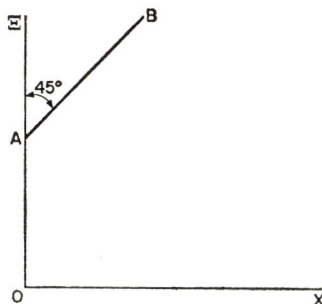

FIG. 1.

which one has to integrate. If we indicate the variable x' on the abscissa axis OX', and the variable ξ on the ordinate axis $O\Xi$, then we obtain the surface over which we have to integrate by making $OA = x$ and extending the line AB to infinity at an angle of 45° with the coordinate axes. The infinite trapezoid $XOAB$ is then the surface over which the integration is to be extended. This latter method of representing the limits has the advantage of perspicuousness.

the limits x and $x+dx$. We have integrated over all other variables. All the other variables are therefore subject to no restrictive conditions. Hence, $\int dn$ is just the number of collisions in unit volume during time τ before which the kinetic energy of one molecule lies between x and $x+dx$. By each such collision a molecule loses this kinetic energy, so that the collision diminishes the number of molecules whose kinetic energy is between x and $x+dx$ by one.† During time τ there will be in all $\int dn$ of such collisions in unit volume. Hence, this number will be decreased by $\int dn$. The number of molecules in unit volume whose kinetic energy at time t lies between x and $x+dx$ is, however, as we know, $f(x, t)dx$; during time τ, this number will decrease by $\int dn$ as a result of the collisions just considered, so that we must subtract $\int dn$ from $f(x, t)dx$.

Up to now we have only considered the collisions by which a molecule loses a kinetic energy between x and $x+dx$, so that $f(x, t)dx$ decreases. We still have to consider those collisions by which a molecule gains this kinetic energy, so that $f(x, t)dx$ thereby increases. If we denote the number of these latter collisions by $\int dv$, then $\int dv$ must be added to $f(x, t)dx$; in the sum

$$f(x, t)dx - \int dn + \int dv \qquad (4)$$

the first term is the number of molecules in unit volume whose kinetic energy is between x and $x+dx$ at time t; from this is subtracted the number of molecules that lose this kinetic energy during time τ, and to it is added the number of molecules that gain this kinetic energy during time τ. The result is clearly the number of molecules that have this kinetic energy at time $t+\tau$, i.e. $f(x, t+\tau)dx$. We thus obtain

$$f(x, t+\tau)dx = f(x, t)dx - \int dn + \int dv \qquad (5)$$

† We have excluded here those collisions for which, after the collision, the kinetic energy of one or even both molecules lies between x and $x+dx$. One sees easily that the number of such collisions—as well as the number of those before which the kinetic energies of both molecules lie between x and $x+dx$, so that two molecules simultaneously lose this kinetic energy—is a higher order infinitesimal and can be neglected. The former collisions, which we are now subtracting incorrectly, are moreover contained in $\int dv$ and hence will be added in again anyway.

We still have to determine $\int dv$. $\int dv$ is the number of collisions in unit volume during time τ, after which the kinetic energy of a molecule lies between x and $x + dx$. We must therefore choose some other notation for the kinetic energies before the collision. Therefore let dv be the number of collisions which occur in unit volume during time τ such that before the collision the kinetic energy of one molecule lies between u and $u + du$, that of the other lies between v and $v + dv$, and after the collision the kinetic energy of one molecule lies between x and $x + dx$. The kinetic energy of the other molecule after the collision is of course determined. dv is therefore the number of collisions which, corresponding to the scheme previously denoted by (A), are characterized by the following scheme:

	a	b	
Before the collision	$u, u + du$	$v, v + dv$	(B)
After the collision	$x, x + dx$		

One sees at once that the collisions characterized by (B) differ from those characterized by (A) only by having different notation for the kinetic energies before and after the collision. The number dv of the latter kind can therefore be found directly from the expression for dn by simply permuting the labels of the variables. Thus we have to write:

u instead of x, v instead of x', x instead of ξ,

and likewise (C)

du instead of dx, dv instead of dx', dx instead of $d\xi$

The number of collisions of the type considered previously was called dn and was given by equation (2). If we carry out the permutation (C), we obtain dv. Hence,

$$dv = \tau . f(u, t) du f(v, t) dv dx \psi(u, v, x)$$

Here we again keep x constant and integrate u and v over all possible values of these quantities. The result,

$$\tau dx \iint f(u, t) f(v, t) \psi(u, v, x) du dv$$

is the number of collisions in unit volume during time τ after which the kinetic energy of one molecule lies between x and $x + dx$. Thus it is the number of collisions by which one molecule gains a kinetic energy between x and $x + dx$, in other words, it is just the number which we earlier called $\int dv$.† Thus, we have

$$\int dv = \tau \,.\, dx \iint f(u, t) f(v, t) \psi(u, v, x) du\, dv \qquad (6)$$

The question of the limits of the double integrals now arises.‡ When $u > x$, v can run through all possible values from zero to

---

† One might think that we have forgotten the collisions after which the kinetic energy of the second of the colliding molecules lies between x and $x + dx$. For such a collision, let $u = u_1$, $v = v_1$. Since we integrate u and v over all possible values, we have also included the collision for which $u = u_1$ and $v_1 = v_1$ and the kinetic energy of the first molecule lies between x and $x + dx$ in the integral; but this is just the case that we were afraid that we had forgotten. For it makes no difference which molecule we call the first and which the second. All these collisions are therefore taken account of in our integral, if we simply write u in place of v and conversely. If one wanted to add a second integral containing collisions after which the kinetic energy of the second molecule lies between x and $x + dx$, then he would have to take the values of u and v without permutation—i.e. integrate v from zero (or $x - u$, resp.) to u, then u from zero to infinity. Only the case where the kinetic energies of both molecules lie between x and $x + dx$ after the collision has not been counted twice, as it should have been; but this is not an error, since the number of such cases is an infinitesimal of higher order.

‡ If we determine the limits according to the method indicated in the footnote on page 97, we obtain the inequalities

$$u \geqq 0, \; v \geqq 0, \; u + v - x \geqq 0$$

If we introduce now some arbitrary new variables p, q, then as is well known

$$dp\, dq = \sum \pm \frac{dp}{dx} \frac{dq}{dv} du\, dv$$

In the special case that we set $p = u + v - x, q = u$, the functional determinant is equal to one (it is of course to be taken positive); furthermore, in this case,

$$v = p + x - q$$

Hence, equation (6) becomes

$$\int dv = \tau \, dx \iint f(q, t) f(p + x - q, t) \psi (q, p + x - q, x) \, dp\, dq$$

The inequalities that determine the limits become

$$q \geqq 0, \; p + x - q \geqq 0, \; p \geqq 0$$

(Footnote continued on next page

infinity; but if $u > x$, then v cannot be smaller than $x - u$, since otherwise $u + v - x$, which is the kinetic energy of the two molecules after the collision, would be negative. Therefore if $u < x$, v runs through all values from $x - u$ to infinity. The u integral must therefore be broken into two parts, the first from zero to x, the second from x to infinity. In the first integral one has to integrate over v from $x - u$ to infinity, and in the second from zero to infinity. The formula (6) thus becomes, when one determines the limits correctly,

$$\int dv = \tau dx \int_0^x \int_{x-u}^{\infty} f(u, t) f(v, t) \psi(u, v, x) du dv$$

$$+ \tau dx \int_x^{\infty} \int_0^{\infty} f(u, t) f(v, t) \psi(u, v, x) du dv \qquad (7)$$

We now introduce in place of v the new variable

$$w = u + v - x \qquad (8)$$

so that we have $v = x + w - u$. Since in the v integration, u as well as x is to be considered constant, it follows from the formula (8) that $dw = dv$. Hence, using the correct determination of limits for the w integration,

$$\int dv = \tau \, dx \int_0^x \int_0^{\infty} f(u, t) f(x+w-u, t) \, \psi(u, x+w-u, x) \, du dw$$

$$+ \tau dx \int_x^{\infty} \int_{u-x}^{\infty} f(u, t) f(x+w-u, t) \psi(u, x+w-u, x) du dv \qquad (9)$$

We can now label the variables in the integral just as we please if we also use the same labels in the inequalities. If we change the letters p, q to x', ξ, then we obtain

$$\int dv = \tau \, dx \iint f(\xi, t) f(x+x'-\xi, t) \psi (\xi, x+x'-\xi, x) dx' \, d\xi$$

The limits are determined by

$$x' \geqq 0, \, \xi \geqq 0, \, x+x'-\xi \geqq 0$$

thus we have come back to the inequalities (3a) which determine the limits in equation (3). If we attach the appropriate limits to the integral signs the last formula agrees with equation (11) in the text, which we have thus obtained here by a quicker method.

Since these integrals represent just a summation of a number of collisions, we can invert the order of integration without difficulty. The first double integral in formula (9) then becomes:

$$\int_0^\infty \int_0^x f(u, t)f(x+w-u, t)\psi(u, x+w-u, x)dwdu \qquad (10)$$

For the second integral, the determination of the new limits of integration is not quite so simple. We shall do it by geometric considerations (Fig. 2). We indicate the value of u on the abscissa axis OU, and the value of w on the ordinate axis OW. x is constant during this integration. We make $OA = x$ and draw through A

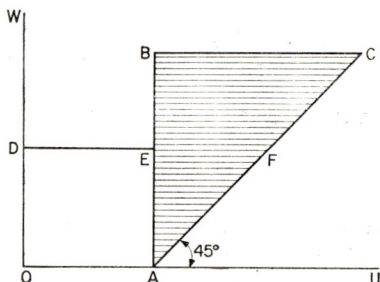

FIG. 2.

the two infinite lines AB parallel to OW, and AC inclined at a 45° angle to the coordinate axes. In the second double integral of (9) we have to integrate over u from x to infinity, i.e. from the point A to infinity; we have to integrate over w from $u-x$ to infinity, i.e. from the line AC to infinity. The total integration is thus to be extended over the unbounded triangle which is shaded in the figure. It is now easy to determine the limits if we want to integrate first over u and then over w. For a given w, say for example $w = OD$, we want to integrate u from DE to DF, hence from x to $x+w$. Thus the w integration goes from zero to infinity. The second double integral in (9) therefore becomes

$$\int_0^\infty \int_x^{x+w} f(u, t)f(x+w-u, t)\psi(u, x+w-u, x)dwdu$$

This can now be combined with the first term, given in (10), into a single double integral. (The first term, incidentally, represents the integration over the infinite rectangle *WOAB* in our figure.) The reunification of the two double integrals gives:

$$\int dv = \tau dx \int_0^\infty \int_0^{x+w} f(u, t)f(x+w-u, t)\psi(u, x+w-u, x)dwdu$$

In order to make our expression uniform with that given by formula (3) for $\int dn$, I will replace w by x' and u by ξ. As is well known, in a definite integral the variables that are integrated over can be denoted however one likes, as long as the limits remain the same. Thereby we obtain

$$\int dv = \tau dx \int_0^\infty \int_0^{x+x'} f(\xi, t)f(x+x'-\xi, t)\psi(\xi, x+x'-\xi, x)dx'd\xi \tag{11}$$

Before we substitute the two values found for $\int dn$ and $\int dv$ into equation (5), we shall transform this equation again. We expand its left-hand side according to Taylor's theorem:

$$f(x, t)dx+\frac{\partial f(x, t)}{\partial t}\tau dx+A\tau^2 dx = f(x, t)dx-\int dn+\int dv$$

where A is some finite quantity; hence

$$\frac{\partial f(x, t)}{\partial t} = \frac{\int dv}{\tau dx}-\frac{\int dn}{\tau dx}-A\tau$$

so that on substituting the values (3) and (11) for $\int dn$ and $\int dv$ we obtain

$$\frac{\partial f(x, t)}{\partial t} = \int_0^\infty \int_0^{x+x'} f(\xi, t)f(x+x'-\xi, t)\psi(\xi, x+x'-\xi, x)dx'd\xi$$

$$-\int_0^\infty \int_0^{x+x'} f(x, t)f(x', t)\psi(x, x', \xi)dx'd\xi-A\tau$$

Since everything except $A\tau$ is finite, the latter can be neglected. Furthermore, the two integrals can be combined into one, since the variables of integration and the limits are the same in both. Thus we obtain:

$$\frac{\partial f(x, t)}{\partial t} = \int_0^\infty \int_0^{x+x'} [f(\xi, t)f(x+x'-\xi, t)\psi(\xi, x+x'-\xi, x)$$
$$-f(x, t)f(x', t)\psi(x, x', \xi)]dx'd\xi \quad (12)$$

This is the desired partial differential equation, which determines the law of variation of the function f. It needs one more transformation, for which we use the properties of the function ψ expressed by the following two equations, valid for any x, x' and ξ:

$$\psi(x, x', \xi) = \psi(x', x, x+x'-\xi) \quad (13)$$

$$\sqrt{xx'}\psi(x, x', \xi) = \sqrt{\xi(x+x'-\xi)}\psi(\xi, x+x'-\xi, x) \quad (14)$$

where of course all the roots are to be taken with the positive sign; the ψ's are also essentially positive quantities. The first of these two equations can easily be proved. Let dn' be the number of collisions that take place in unit volume during the very small time interval (previously denoted by τ) in such a way that before the collision the kinetic energy of the first molecules lies between x' and $x'+dx'$, and that of the second between x and $x+dx$; and after the collision, the kinetic energy of the first molecule lies between $x+x'-\xi-d\xi$ and $x+x'-\xi$. The collision is characterized by the scheme:

	a	b	
Before the collision	$x', x'+dx'$	$x, x+dx$	(D)
After the collision	$x+x'-\xi-d\xi, x+x'-\xi$		

Then dn' can be found by a simple permutation of labels from the quantity earlier denoted by dn. Indeed, on comparing the schemes (D) and (A), we see that we must write:

x' instead of x, x instead of x', $x+x'-\xi-d\xi$ instead of ξ,

dx' instead of dx, dx instead of dx'.

If one makes these exchanges in equation (2), he obtains:

$$dn' = \tau f(x', t)dx'.f(x, t)dx.d\xi.\psi(x', x, x+x'-\xi-d\xi) \quad (15)$$

However, if the kinetic energy of one molecule after the collision lies between $x+x'-\xi-d\xi$ and $x+x'-\xi$, then that of the other lies between ξ and $\xi+d\xi$. Instead of the scheme (D) we can therefore characterize our collision by the following:

	a	b
Before the collision	$x', x'+dx'$	$x, x+dx$
After the collision		$\xi, \xi+d\xi.$

And now one sees that this is exactly the same collision as the one characterized by the scheme (A). For clearly it makes no difference which molecule I call the first (a) and which the second (b). Since the two kinds of collisions are the same, their numbers must be equal, hence $dn = dn'$. If we equate the two values (2) and (15) and cancel the common factors, we obtain

$$\psi(x, x', \xi) = \psi(x', x, x+x'-\xi-d\xi)$$

Here the differential $d\xi$ can be neglected compared to finite quantities, since ψ cannot be discontinuous, and thus we obtain equation (13).

The proof of equation (14) is more difficult. The proof was first given, though in a somewhat different form, by Maxwell; it was then considerably generalized by me, and was shown to be a special case of Jacobi's principle of the last multiplier; hence, I do not have to give the proof here, but can assume it to be known. I remark only that in the proof it is assumed that the force between two material points is a function of their distance, which acts in the direction of their line of centres, and that action and reaction are equal. These assumptions are also necessary for the validity of the following calculations. Taking account of (14), we can take a common factor ψ out of the brackets in equation (12) and obtain:

$$\frac{\partial f(x, t)}{\partial t} = \int_0^\infty \int_0^{x+x'} \left[\frac{f(\xi, t)}{\sqrt{\xi}} \frac{f(x+x'-\xi, t)}{\sqrt{x+x'-\xi}} - \frac{f(x, t)}{\sqrt{x}} \frac{f(x', t)}{\sqrt{x'}} \right]$$
$$\times \sqrt{xx'}\psi(x, x', \xi)dx'd\xi \quad (16)$$

This is the fundamental equation for the variation of the function $f(x, t)$. I remark again that the roots are to be taken positive, and that both ψ and f are essentially positive quantities. If for a moment we set

$$f(x, t) = C\sqrt{x}e^{-hx} \tag{16a}$$

where C and h are constants, so that therefore

$$f(x', t) = C\sqrt{x'}e^{-hx'}, \quad f(\xi, t) = C\sqrt{\xi}e^{-h\xi},$$
$$f(x+x'-\xi, t) = C\sqrt{x+x'-\xi}e^{-h(x+x'-\xi)}$$

then the expression in brackets in equation (16) vanishes; hence $\partial f(x, t)/\partial t = 0$. This is nothing more than Maxwell's proof transcribed in our present notation. If the distribution of states at any time is determined by the formula (16a), that $\partial f(x, t)/\partial t = 0$, i.e. this distribution does not change further in the course of time. This, and nothing else, is what Maxwell proved. However, we shall now consider the problem more generally. We shall assume that the distribution of kinetic energy is initially a completely arbitrary one, and we shall ask ourselves how it changes in the course of time. Its variation is determined by the partial differential equation (16). This partial differential equation can be transformed, as we shall see later on, into a system of ordinary differential equations, if one replaces the double integrals by a sum of many terms. Indeed, as is well known, such a double integral is nothing more than an abbreviated notation for a sum of infinitely many terms. The methods of calculation are then more intuitively clear for the system of ordinary differential equations. However, I will not at first undertake this exchange of summation and integration, in order that it will not appear to be necessary for the proof of our theorem. This proof can be carried out without abandoning the symbolism of integral calculus. It is only for the sake of better understanding that we use the summation formulae at the end.

We shall first give the proof of a theorem which forms the basis of our present investigation: the theorem that the quantity

$$E = \int_0^\infty f(x, t) \left\{ \log\left[\frac{f(x, t)}{\sqrt{x}}\right] - 1 \right\} dx \tag{17}$$

can never increase, when the function $f(x, t)$ that occurs in the definite integral satisfies the differential equation (16). On the right-hand side of equation (17) one has to integrate x from zero to infinity. Therefore E does not depend on x, but only on t. Since t does not appear in the limits of the integral, we can obtain the time derivative dE/dt by finding the partial derivative with respect to t of the quantity under the integral sign, keeping x constant. This differentiation, which can very easily be carried out, yields

$$\frac{dE}{dt} = \int_0^\infty \log\left[\frac{f(x, t)}{x}\right] \frac{\partial f(x, t)}{\partial t} dx$$

We assume that $f(x, t)$ satisfies equation (16). If we substitute from this equation the value for $\partial f(x, t)/\partial t$, then we obtain

$$\frac{dE}{dt} = \int_0^\infty \log\left[\frac{f(x, t)}{\sqrt{x}}\right] dx \int_0^\infty \int_0^{x+x'} \left[\frac{f(\xi, t)}{\sqrt{\xi}}\frac{f(x+x'-\xi, t)}{\sqrt{x+x'-\xi}}\right.$$

$$\left. -\frac{f(x, t)}{\sqrt{x}}\frac{f(x', t)}{\sqrt{x'}}\right] \times \sqrt{xx'}\psi(x, x', \xi)dx'd\xi$$

Since in the integration over x' and ξ the quantity x is to be kept fixed, we can also put the logarithm under the two following integration signs and write

$$\frac{dE}{dt} = \int_0^\infty \int_0^\infty \int_0^{x+x'} \log\frac{f(x, t)}{\sqrt{x}}\left[\frac{f(\xi, t)f(x+x'-\xi, t)}{\sqrt{\xi}\sqrt{x+x'-\xi}} - \frac{f(x, t)f(x', t)}{\sqrt{x}\sqrt{x'}}\right]$$

$$\times \sqrt{xx'}\psi(x, x', \xi)dxdx'd\xi \qquad (18)$$

The real meaning of the transformation which we are now going to perform with this expression will of course become completely clear only when we replace the integrals by sums. It will then appear that all the following transformations of integrals are nothing more than changes in the order of summation; it will also be clear just why these changes in the order of summation are necessary. However, I will not discuss this any further at this point, but simply proceed as quickly as possible to the proof of the desired theorem, that E cannot increase. In equation (18) we can integrate first over x' and then over x;† we thereby obtain:

$$\frac{dE}{dt} = \int_0^\infty \int_0^\infty \int_0^{x+x'} \log\frac{f(x,t)}{\sqrt{x}} \left[\frac{f(\xi,t)}{\sqrt{\xi}} \frac{(x+x'-\xi,t)}{\sqrt{x+x'-\xi}} - \frac{f(x,t)}{\sqrt{x}}\frac{f(x',t)}{\sqrt{x'}} \right]$$
$$\times \sqrt{xx'}\psi(x,x',\xi)dx'dxd\xi \quad (19)$$

or when we substitute for $\psi(x, x', \xi)$ its value from equation (13),

$$\frac{dE}{dt} = \int_0^\infty \int_0^\infty \int_0^{x+x'} \log\frac{f(x,t)}{\sqrt{x}} \left[\frac{f(\xi,t)f(x+x'-\xi,t)}{\sqrt{\xi}\sqrt{x+x'-\xi}} - \frac{f(x,t)}{\sqrt{x}}\frac{f(x',t)}{\sqrt{x'}} \right]$$
$$\times \sqrt{xx'}\psi(x',x,x+x'-\xi)dx'dxd\xi$$

We now leave the variables x' and x unchanged, but instead of ξ we introduce the new variable $\xi' = x+x'-\xi$, so that $\xi = x+x' - \xi'$, $d\xi = -d\xi'$. Then we have

$$\frac{dE}{dt} = -\int_0^\infty \int_0^\infty \int_{x+x'}^0 \log\frac{f(x,t)}{\sqrt{x}} \left[\frac{(x+x'-\xi',t)f(\xi',t)}{\sqrt{x+x'-\xi'}\sqrt{\xi'}} \right.$$
$$\left. -\frac{f(x,t)}{\sqrt{x}}\frac{f(x',t)}{\sqrt{x'}} \right] \times \sqrt{xx'}\psi(x',x,\xi')dx'dxd\xi'$$

---

† That the change in order of integration is unconditionally allowed follows from the fact that we could have derived equations (20), (22) and (23) directly in just the same way that we have obtained (18); we have adopted the method of transformations merely in order not to have to repeat four times the arguments by which we obtained equation (18). Also, the fact that the earlier differentiation under the integral sign is not permitted, because of the discontinuity of the integrand, does not destroy the proof in the text, as one can show by excluding from the entire space over which the integration extends in equations (18), (20), (22) and (23) a very thin strip on the surface at those places where one of the quantities s, s', σ, or σ' is zero or infinite. One can prove by means of Taylor's theorem that the sum of the terms thus excluded from $4\,dE/dt$ cannot be positive, if none of these quantities is infinite or has infinitely many discontinuities.

or, when we reverse the sign and limits of the first integral,

$$\frac{dE}{dt} = \int_0^\infty \int_0^\infty \int_0^{x+x'} \log\frac{f(x,t)}{\sqrt{x}}\left[\frac{f(\xi',t)f(x+x'-\xi',t)}{\sqrt{\xi}\sqrt{x+x'-\xi'}} - \frac{f(x,t)f(x',t)}{\sqrt{x}\sqrt{x'}}\right]$$
$$\times \sqrt{xx'}\psi(x',x,\xi')dx'dxd\xi$$

This triple integral is now constructed in the same way as that in equation (18); but the variables over which one has to integrate are labelled differently. But this is only an apparent difference. The variables of integration of a definite integral may be labelled in any way one likes, as long as the limits remain the same. Hence, we can write ξ in place of ξ' in (19), and also permute the letters x and x'. Thereby we obtain

$$\frac{dE}{dt} = \int_0^\infty \int_0^\infty \int_0^{x+x'} \log\frac{f(x',t)}{\sqrt{x'}}\left[\frac{f(\xi,t)f(x+x'-\xi,t)}{\sqrt{\xi}\sqrt{x+x'-\xi}} - \frac{f(x,t)f(x',t)}{\sqrt{x}\sqrt{x'}}\right]$$
$$\times \sqrt{xx'}\psi(x,x',\xi)dxdx'd\xi \quad (20)$$

There can be no doubt as to the identity of the two integrals (19) and (20), since they differ only by the labels by which the variables of integration are denoted.

We obtain a third expression for dE/dt in the following way. We substitute in (18) in place of $\sqrt{xx'}\,\psi(x,x',\xi)$ its value from equation (14). Thereby we obtain:

$$\frac{dE}{dt} = \int_0^\infty \int_0^\infty \int_0^{x+x'} \log\frac{f(x,t)}{\sqrt{x}}\left[\frac{f(\xi,t)f(x+x'-\xi,t)}{\sqrt{\xi}\sqrt{x+x'-\xi}} - \frac{f(x,t)f(x',t)}{\sqrt{x}\sqrt{x'}}\right]$$
$$\times \sqrt{\xi(x+x'-\xi)}\psi(\xi,x+x'-\xi,x)dxdx'd\xi$$

We will now introduce a new variable for x'. We want to do the integration over x' first, before the integration over ξ. We need to transform the double integral

$$\int_0^\infty \int_0^{x+x'} \log\frac{f(x,t)}{\sqrt{x}}\left[\frac{f(\xi,t)f(x+x'-\xi,t)}{\sqrt{\xi}\sqrt{x+x'-\xi}} - \frac{f(x,t)}{\sqrt{x}}\frac{f(x',t)}{\sqrt{x'}}\right]$$
$$\times \sqrt{\xi(x+x'-\xi)}\psi(\xi,x+x'-\xi,x)dx'd\xi$$

In order to obtain dE/dt we then simply multiply this by dx and integrate over x from zero to infinity. We have already previously inverted the order of integration in such a double integral. By the same argument as before we can transform it to a sum of two integrals:

$$
\int_0^x \int_0^\infty \log\frac{f(x,t)}{\sqrt{x}}\left[\frac{f(\xi,t)f(x+x'-\xi,t)}{\sqrt{\xi}\ \sqrt{x+x'-\xi'}} - \frac{f(x,t)}{\sqrt{x}}\frac{f(x',t)}{\sqrt{x'}}\right]
$$

$$
\times \sqrt{\xi(x+x'-\xi)}\psi(\xi,x+x'-\xi,x)d\xi dx'
$$

$$
+\int_x^\infty \int_{\zeta-x}^\infty \log\frac{f(x,t)}{\sqrt{x}}\left[\frac{f(\xi,t)f(x+x'-\xi,t)}{\sqrt{\xi}\ \sqrt{x+x'-\xi}} - \frac{f(x,t)}{\sqrt{x}}\frac{f(x',t)}{\sqrt{x'}}\right]
$$

$$
\times \sqrt{\xi(x+x'-\xi)}\psi(\xi,x+x'-\xi,x)d\xi dx'
$$

If we introduce in these two integrals the variable

$$
\xi' = x+x'-\xi
$$

in place of x', we obtain, with a correct determination of the limits:

$$
\int_0^x \int_{x-\xi}^\infty \log\frac{f(x,t)}{\sqrt{x}}\left[\frac{f(\xi,t)f(\xi',t)}{\sqrt{\xi}\ \sqrt{\xi'}} - \frac{f(x,t)}{\sqrt{x}}\frac{f(\xi+\xi'-x,t)}{\sqrt{\xi+\xi'-x}}\right]
$$

$$
\times \sqrt{\xi\xi'}\psi(\xi,\xi',x)d\xi d\xi'
$$

$$
+\int_x^\infty \int_0^\infty \log\frac{f(x,t)}{\sqrt{x}}\left[\frac{f(\xi,t)f(\xi',t)}{\sqrt{\xi}\ \sqrt{\xi'}} - \frac{f(x,t)}{\sqrt{x}}\frac{f(\xi+\xi'-x,t)}{\sqrt{\xi+\xi'-x}}\right]
$$

$$
\times \sqrt{\xi\xi'}\psi(\xi,\xi',x)d\xi d\xi'
$$

These two definite integrals are to be integrated over x from zero to infinity, so that we obtain:

$$\frac{dE}{dt} = \int_0^\infty \int_0^x \int_{x-\xi}^\infty \log\frac{f(x,t)}{\sqrt{x}}\left[\frac{f(\xi,t)}{\sqrt{\xi}}\frac{f(\xi',t)}{\sqrt{\xi'}} - \frac{f(x,t)}{\sqrt{x}}\frac{f(\xi+\xi'-x,t)}{\sqrt{\xi+\xi'-x}}\right]$$

$$\times \sqrt{\xi\xi'}\psi(\xi,\xi',x)dxd\xi d\xi'$$

$$+ \int_0^\infty \int_x^\infty \int_0^\infty \log\frac{f(x,t)}{\sqrt{x}}\left[\frac{f(\xi,t)}{\sqrt{\xi}}\frac{f(\xi',t)}{\sqrt{\xi'}} - \frac{f(x,t)}{\sqrt{x}}\frac{f(\xi+\xi'-x,t)}{\sqrt{\xi+\xi'-x}}\right]$$

$$\times \sqrt{\xi\xi'}\psi(\xi,\xi',x)dxd\xi d\xi' \tag{21}$$

We must now change the order of integration so that we integrate first over x, then over ξ', and last over ξ.† For the purpose of determining limits of integration it is best to visualize the integration space geometrically (Fig. 3). Since the integral is

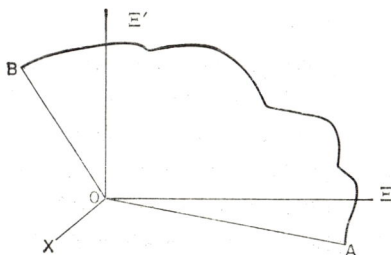

FIG. 3.

a triple one, we must use three-dimensional space. We construct three rectangular coordinate axes OX, $O\Xi$, $O\Xi'$ in space, and indicate on these the values of x, ξ, and ξ'. Further, we draw in the

(*Text continued on page 114*)

---

† All the rather tedious determinations of limits in the text can be very much simplified by using the method already indicated in the footnote on page 97. Then equation (18) reads as follows:

$$\frac{dE}{dt} = \iiint \log\frac{f(x,t)}{\sqrt{x}}\left[\frac{f(\xi,t)}{\sqrt{\xi}}\cdot\frac{f(x+x'-\xi,t)}{\sqrt{x+x'-\xi}} - \frac{f(x,t)}{\sqrt{x}}\frac{f(x',t)}{\sqrt{x'}}\right]$$

$$\times \sqrt{xx'}\psi(x,x'\ \xi)\,dx\,dx'\,d\xi \tag{18a}$$

(*Footnote continued on next page*)

where the integration is over all values satisfying the inequalities:

$$x \geqq 0, \ x' \geqq 0, \ \xi \geqq 0, \ x+x'-\xi \geqq 0 \tag{18b}$$

The two equations (18a) and (18b) are equivalent to the single definite integral (18), and I remark that now the order of integration is completely arbitrary; there is no requirement that we must integrate first over one variable and then another, but instead we just have to integrate over all values that satisfy (18b). If we now introduce some arbitrary new variables u, v, w, then as is well known

$$du \, dv \, dw = \sum_{\iota} \pm \frac{du}{dx} \frac{dv}{dy} \frac{dw}{dz} \, dx \, dy \, dz$$

If we now wish to obtain equation (20) of the text, then we need only set

$$u = x', \ v = x, \ w = x+x'-\xi$$

Then the functional determinant is equal to one, and it is clear that it is to be taken with the positive sign, if we always integrate from smaller to larger values of the variables, so that the differentials are to be considered positive. Then

$$du \, dv \, dw = dx \, dx' \, d\xi$$

and equation (18a) becomes

$$\frac{dE}{dt} = \iiint \log \frac{f(v,t)}{\sqrt{v}} \left[\frac{f(u+v-w,t)}{\sqrt{u+v-w}} \frac{f(w,t)}{\sqrt{w}} - \frac{f(v,t)}{\sqrt{v}} \frac{f(u,t)}{\sqrt{u}} \right]$$
$$\times \sqrt{uv} \psi \, (v, u, u+v-w) \, du \, dv \, dw$$

The inequalities (18b), which determine the limits, become

$$v \geqq 0, \ u \geqq 0, \ u+v-w \geqq 0, \ w \geqq 0$$

We can now again replace the letters u, v, w by x, x' and ξ (since the labels of integration variables are arbitrary) and obtain for the integral

$$\frac{dE}{dt} = \iiint \log \frac{f(x',t)}{\sqrt{x'}} \left[\frac{f(x+x'-\xi,t)}{\sqrt{x+x'-\xi}} \frac{f(\xi,t)}{\sqrt{\xi}} - \frac{f(x,t)}{\sqrt{x}} \frac{f(x',t)}{\sqrt{x'}} \right]$$
$$\times \sqrt{xx'} \psi \, (x', x, x+x'-\xi) \, dx \, dx' \, d\xi$$

and for the inequalities, which determine the limits,

$$x \geqq 0, \ x' \geqq 0, \ \xi \geqq 0, \ x+x'-\xi \geqq 0 \tag{20b}$$

If we now finally replace $\psi(x', x, x+x'-\xi)$ by $\psi(x, x', \xi)$ using equation (13), we obtain

$$\frac{dE}{dt} = \iiint \log \frac{f(x',t)}{\sqrt{x'}} \left[\frac{f(\xi,t)}{\sqrt{\xi}} \frac{f(x+x'-\xi,t)}{\sqrt{x+x'-\xi}} - \frac{f(x,t)}{\sqrt{x}} \frac{f(x',t)}{\sqrt{x'}} \right]$$
$$\times \sqrt{xx'} \, \psi(x, x', \xi) \, dx \, dx' \, d\xi \tag{20a}$$

The inequalities (20b) are identical with the inequalities (18a). If we combine the two equations (20a) and (20b) into a single one, by supposing that we integrate first over ξ, then over x', and last over x, and that the limits of integration are attached to the integral signs, then we obtain the desired equation (20) in the text.

If we want to obtain equations (22) of the text, then we set

$$u = \xi, v = x+x'-\xi, w = x$$

The determinant is again one, hence

$$du\, dv\, dw = dx\, dx'\, d\xi$$

Furthermore,

$$x' = u+v-w$$

Equation (18a) becomes, after introducing these variables:

$$\frac{dE}{dt} = \iiint \log \frac{f(w,t)}{\sqrt{w}} \left[\frac{f(u,t)}{\sqrt{u}} \frac{f(v,t)}{\sqrt{v}} - \frac{f(w,t)}{\sqrt{w}} \frac{f(u+v-w,t)}{\sqrt{u+v-w}} \right]$$

$$\times \sqrt{w(u+v-w)}\ \psi(w, u+v-w, u)\, du\, dv\, dw$$

and the inequalities (18b) become

$$w \geqq 0,\ u+v-w \geqq 0,\ u \geqq 0,\ v \geqq 0$$

If, as before, we replace the letters u, v, w by x, x', ξ, then we obtain

$$\frac{dE}{dt} = \iiint \log \frac{f(\xi, t]}{\sqrt{\xi}} \left[\frac{f(x, t]}{\sqrt{x}} \frac{f(x', t]}{\sqrt{x'}} - \frac{f(\xi, t]}{\sqrt{\xi}} \frac{f(x+x'-\xi, t]}{\sqrt{x+x'-\xi}} \right]$$

$$\times \sqrt{\xi(x+x'-\xi)}\ \psi(\xi, x+x'-\xi, x)\, dx\, dx'\, d\xi \qquad (22a)$$

$$x \geqq 0,\ x' \geqq 0,\ x+x'-\xi \geqq 0,\ \xi \geqq 0$$

We see that we must simply use (14) and write in the limits of the integrals in order to obtain equation (22) of the text from (22a). One sees therefore that if the method of determination of limits by inequalities is used, the transformations can be made with hardly any calculation at all although detailed calculations were needed in the text. I used the longer method in the text simply because the method of inequalities is somewhat unusual.

I note here that one could also use the following expression for E:

$$E_1 = \int_0^\infty f(x,t) \log \left[\frac{f(x,t)}{\sqrt{x}} \right] dx \qquad (17a)$$

This expression is different from the one given in the text for E simply by a term $\int_0^\infty f(x,t)\, dx$, which is just the total number of molecules in unit volume.

Since this total number is constant, we see that E_1 is also a quantity that cannot increase.

Another transformation method could be developed by defining the function $\psi(x, x', \xi)$ to be zero whenever $x+x'-\xi < 0$; all the integrations could then be extended from zero to infinity.

plane $XO\Xi$ the line OA, which makes 45° angles with OX and $O\Xi$, and likewise in the plane $\Xi O\Xi'$ the line OB. We now consider the first triple integral in (21). Here we have to integrate ξ' from $x-\xi$ to infinity, i.e. from a point of the plane AOB to infinity. We have to integrate ξ from zero to x, i.e. from zero to a point of the line OA. The integration space of the first integral is therefore just that part of space which stands vertically on AOB (if one imagines that the $O\Xi'$ axis is vertical). Likewise one finds that the integration space of the second integral in (21) is the part of space that stands vertically on the triangle $AO\Xi$. (The triangle formed by extending the lines OA and $O\Xi$ to infinity.) The two integrals can therefore be represented as a single integration over a solid angle bounded by the four figures AOB, $AO\Xi$, $BO\Xi'$ and $\Xi O\Xi'$. Now it is easy to determine the limits if we first integrate over x. For constant ξ and ξ' we remain in the solid angle as x goes from zero to $\xi+\xi'$. Zero and $\xi+\xi'$ are therefore the limits of integration for x. The integration over ξ and ξ' simply goes from zero to infinity. With this new order of integration we can combine the two integrals into one, and we have

$$\frac{dE}{dt} = \int_0^\infty \int_0^\infty \int_0^{\xi+\xi'} \log\frac{f(x,t)}{\sqrt{x}} \left[\frac{f(\xi,t)}{\sqrt{\xi}}\frac{f(\xi',t)}{\sqrt{\xi'}} \right.$$
$$\left. -\frac{f(x,t)}{\sqrt{x}}\frac{f(\xi+\xi'-x,t)}{\sqrt{\xi+\xi'-x}}\right] \times \sqrt{\xi\xi'}\psi(\xi,\xi',x)d\xi d\xi' dx$$

In this definite integral it again makes no difference how we label the variables over which we integrate. Hence, we can denote the two variables ξ and ξ' by the Latin letters x and x', and the variable x with the letter ξ. If we do this, and put a minus sign in front of the whole integral while reversing the signs inside the bracket, we obtain

$$\frac{dE}{dt} = -\int_0^\infty \int_0^\infty \int_0^{x+x'} \log\frac{f(\xi,t)}{\sqrt{\xi}} \left[\frac{f(\xi,t)}{\sqrt{\xi}}\frac{f(x+x'-\xi,t)}{\sqrt{x+x'-\xi}}-\frac{f(x,t)}{\sqrt{x}}\frac{f(x',t)}{\sqrt{x'}}\right]$$
$$\times \sqrt{xx'}\psi(x,x',\xi)dxdx'd\xi \quad (22)$$

If we apply to this equation the same transformation by which we obtained equation (20) from equation (18), then we obtain still a fourth expression for dE/dt. I do not need to go through this transformation in detail; one sees easily that its result is the following:

$$\frac{dE}{dt} = -\int_0^\infty \int_0^\infty \int_0^{x+x'} \log \frac{f(x+x'-\xi,t)}{\sqrt{x+x'-\xi}} \left[\frac{f(\xi,t)}{\sqrt{\xi}} \frac{f(x+x'-\xi,t)}{\sqrt{x+x'-\xi}} \right.$$
$$\left. -\frac{f(x,t)f(x',t)}{\sqrt{x}\;\sqrt{x'}} \right] \times \sqrt{xx'}\,\psi(x,x',\xi)dxdx'd\xi \tag{23}$$

I will now collect together the four expressions that we have obtained for dE/dt, for which purpose I will use the following abbreviations. I set:

$$\frac{f(x,t)}{\sqrt{x}} = s, \; \frac{f(x',t)}{\sqrt{x'}} = s', \; \frac{f(\xi,t)}{\sqrt{\xi}} = \sigma, \; \frac{f(x+x'-\xi,t)}{\sqrt{x+x'-\xi}} = \sigma'$$

$$\sqrt{xx'}\,\psi(x,x',\xi) = r$$

The four equations (18), (20), (22), and (23) thus reduce to

$$\frac{dE}{dt} = \int_0^\infty \int_0^\infty \int_0^{x+x'} \log s(\sigma\sigma' - ss')rdxdx'd\xi$$

$$\frac{dE}{dt} = \int_0^\infty \int_0^\infty \int_0^{x+x'} \log s'(\sigma\sigma' - ss')rdxdx'd\xi$$

$$\frac{dE}{dt} = -\int_0^\infty \int_0^\infty \int_0^{x+x'} \log \sigma(\sigma\sigma' - ss')rdxdx'd\xi$$

$$\frac{dE}{dt} = -\int_0^\infty \int_0^\infty \int_0^{x+x'} \log \sigma'(\sigma\sigma' - ss')rdxdx'd\xi$$

We also obtain dE/dt when we add together all four expressions and divide by 4. Since on the right there are simply definite integrals with the same integration variables and the same limits, we can write the integral sign in front of the sum and then add the various quantities under the integral sign. Taking out the common factor, we obtain

$$\frac{dE}{dt} = \tfrac{1}{4}\int_0^\infty \int_0^\infty \int_0^{x+x'} (\log s + \log s' - \log \sigma - \log \sigma')(\sigma\sigma' - ss')r\,dx\,dx'\,d\xi$$

or, after replacing the sum of logarithms by a logarithm of a product,

$$\frac{dE}{dt} = \tfrac{1}{4}\int_0^\infty \int_0^\infty \int_0^{x+x'} \log\left(\frac{ss'}{\sigma\sigma'}\right)(\sigma\sigma' - ss')r\,dx\,dx'\,d\xi \qquad (24)$$

If it is not true for all combinations of values of the variables contained in the s and σ that

$$ss' = \sigma\sigma' \qquad (25)$$

then for some combinations we must have either $ss' > \sigma\sigma'$ or $ss' < \sigma\sigma'$. In the first case, $\log(ss'/\sigma\sigma')$ is positive but $\sigma\sigma' - ss'$ is negative, and in the second case the converse is true; in both cases the product $\log(ss'/\sigma\sigma')(\sigma\sigma' - ss')$ is negative. Now the quantity r is essentially positive, since ψ is always positive, and the square roots are to be taken with positive signs. Hence, the quantity under the integral sign, and therefore the entire integral, is necessarily negative. Therefore E must necessarily decrease. Only when equation (25) holds can E remain constant. Since, as we shall see later on, E cannot go to negative infinity, it must approach a minimum with increasing time, and at the minimum we have $dE/dt = 0$, hence equation (25) holds. This equation implies, when we substitute for s, s', σ, and σ' their definitions,

$$\frac{f(x,t)f(x',t)}{\sqrt{x}\;\sqrt{x'}} = \frac{f(\xi,t)f(x+x'-\xi,t)}{\sqrt{\xi}\;\sqrt{x+x'-\xi}}$$

In order that this equation may be satisfied for all values of the variables x, x', and ξ, it is necessary, as can easily be shown, that

$$f(x,t) = C\sqrt{x}\,e^{-hx}$$

It has thus been rigorously proved that, whatever may be the initial distribution of kinetic energy, in the course of a very long time it must always necessarily approach the one found by Maxwell. The procedure used so far is of course nothing more than a mathematical artifice employed in order to give a rigorous proof of a theorem whose exact proof has not previously been found. It gains meaning by its applicability to the theory of polyatomic gas molecules. There one can again prove that a certain quantity E can only decrease as a consequence of molecular motion, or in a limiting case can remain constant. One can also prove that for the atomic motion of a system of arbitrarily many material points there always exists a certain quantity which, in consequence of any atomic motion, cannot increase, and this quantity agrees up to a constant factor with the value found for the well-known integral $\int dQ/T$ in my paper on the " Analytical proof of the 2nd law, etc."† We have therefore prepared the way for an analytical proof of the second law in a completely different way from those previously investigated. Up to now the object has been to show that $\int dQ/T = 0$ for reversible cyclic processes, but it has not been proved analytically that this quantity is always negative for irreversible processes, which are the only ones that occur in nature. The reversible cyclic process is only an ideal, which one can more or less closely approach but never completely attain. Here, however, we have succeeded in showing that $\int dQ/T$ is in general negative, and is equal to zero only for the limiting case, which is of course the reversible cyclic process (since if one can go through the process in either direction, $\int dQ/T$ cannot be negative).

II. Replacement of Integrals by Sums

I will not consider any further the relation between E and $\int dQ/T$, but rather now show how all the previous calculations can be made much clearer when the partial differential equation (16) is transformed to a system of ordinary differential equations. This is done

---

† L. Boltzmann, *Wien. Ber.* **63**, 712 (1871).

E

by replacing the double integral which appears in the partial differential equation by a sum, according to the well-known formula

$$\int_0^\infty f(x,t)dx = \lim \varepsilon[f(\varepsilon,t)+f(2\varepsilon,t)+f(3\varepsilon,t)+\ldots f(p\varepsilon,t)]$$

$$\text{for } \lim \varepsilon = 0, \quad \lim p\varepsilon = \infty$$

We wish to replace both integrals in (16) by such sums, and first take ε and p both finite. The (16) becomes a differential equation with the following unknowns: $f(\varepsilon,t), f(2\varepsilon,t),\ldots,f(p\varepsilon,t)$. Each unknown is a function only of time. The number of unknowns is p. But equation (16) must hold for each x. If we substitute therein the values

$$x = \varepsilon, \ x = 2\varepsilon, \ldots, \ x = p\varepsilon$$

then we obtain in all p differential equations relating our p unknowns; and since the unknowns are functions only of time, the differential equations are not partial ones. This system of p ordinary differential equations in p unknowns will first be solved, and then we shall investigate the limit approached by the solution when ε becomes infinitesimal and $p\varepsilon$ becomes infinite. The limit is then the solution of the partial differential equation. The substitution of summation formulae into the partial differential equation offers no difficulty. It is transformed into the system of equations (34) below. We shall carry out explicitly the calculations which are only sketched here. I will also show how one must modify our problem in order to arrive directly at the system of p ordinary differential equations instead of first deriving a partial differential equation. The method that we shall use is by no means new. As is well known, the integral is nothing more than a symbolic notation for the sum of infinitely many infinitesimal terms. The symbolic notation of the integral calculus has the advantage of such great brevity that in most cases it would only lead to useless complications to write out the integral first as a sum of p terms and then let p become large. In spite of this, there are cases in which the latter method—on account of its generality, and especially on account of its greater perspicuousness, in that it allows the various solutions of a problem to appear—should not be completely rejected. I recall the elegant solution of the problem of string-vibrations by Lagrange in the *Miscellanea taurinensia*, where he first treated the vibrations of a system of *n* spheres bound together, and then obtained the vibrations of a string by letting *n* become large while the mass of each sphere became small.† The problem of diffusion and heat conduction was solved in a similar way by Stefan.‡ Another beautiful application of this method was made by Reimann for the differential equation§

$$\frac{d^2w}{dr\,ds} = a\left(\frac{dw}{dr} + \frac{dw}{ds}\right)$$

It appears to me that in our case also, once one has become used to some abstractions, this method very much improves clarity. We wish to replace the continuous variable x by a series of discrete values $\varepsilon, 2\varepsilon, 3\varepsilon, \ldots p\varepsilon$. Hence we must assume that our molecules are not able to take up a continuous series of kinetic energy values, but rather only values that are multiples of a certain quantity ε. Otherwise we shall treat exactly the same problem as before. We have many gas molecules in a space R. They are able to have only the following kinetic energies:

$$\varepsilon, \ 2\varepsilon, \ 3\varepsilon, \ 4\varepsilon, \ldots \ p\varepsilon \qquad (26)$$

No molecule may have an intermediate or a greater kinetic energy. When two molecules collide, they can change their kinetic energies in many different ways. However, after the collision the kinetic energy of each molecule must always be a multiple of ε. I certainly do not need to remark that for the moment we are not concerned with a real physical problem. It would be difficult to imagine an apparatus that could regulate the collisions of two bodies in such a way that their kinetic energies after a collision are always

---

† [J. L. LaGrange, *Misc. Taur.* **1**, (1759), reprinted in his *Oeuvres*, Gauthier-Villars, Paris, 1867, **1**, 37.]

‡ J. Stefan, *Wien. Ber.* **47**, 327 (1863), and R. Beez [*Z. Math. Phys.* **7**, 327 (1862), **10**, 358 (1865)].

§ B. Riemann, *Abh. K. Ges. Wiss. Göttingen* **8**, 43 (1859).

multiples of ε. That is not the question here. In any case we are free to study the mathematical consequences of this assumption, which is nothing more than an artifice to help us to calculate physical processes. For at the end we shall make ε infinitely small and $p\varepsilon$ infinitely large, so that the series of kinetic energies given in (26) will become a continuous one, and our mathematical fiction will reduce to the physical problem treated earlier.

We now assume that at time t there are w_1 molecules with kinetic energy ε, w_2 with kinetic energy 2ε, ... and w_p with kinetic energy $p\varepsilon$, in unit volume. We again assume that the distribution of kinetic energy at time t is already uniform (the w's are independent of position in space) and all directions of velocity are equally probable. In the course of time, molecules with a certain kinetic energy, for example, $k\varepsilon$, will leave unit volume; but since the distribution of kinetic energy is uniform, on the average the same amount will come in from the surroundings. Since it is only a question of average values, the w's will change only by collisions. If we want to establish a differential equation for the changes of the w's, we must subject the collisions to a closer examination. We denote by $N_{\kappa\lambda}^{kl}$ the number of collisions in unit volume that occur during a very small time τ, such that the kinetic energies of the two molecules are $k\varepsilon$ and $l\varepsilon$ before the collision, and $\kappa\varepsilon$ and $\lambda\varepsilon$ after the collision. The four quantities k, l, κ, λ are positive integers, $\leqq p$; for collisions in which k, l, κ, λ have other values do not occur. Moreover we have the equation

$$k+l = \kappa+\lambda \tag{27}$$

since the sum of the kinetic energies of the two molecules must be the same before and after the collision. Since we are not at present dealing with a real physical problem, we cannot actually determine these numbers $N_{\kappa\lambda}^{kl}$; we can only make some rather arbitrary assumptions about them and study the consequences thereof. However, if we want our problem to reduce to the one treated earlier in the limit of infinitely small ε, then we must assume that the $N_{\kappa\lambda}^{kl}$ are determined in just the same way as the collision numbers previously were. We therefore assume that $N_{\kappa\lambda}^{kl}$ is proportional to the time τ, to the number of molecules with kinetic energy $k\varepsilon$ in unit volume, w_k, and to w_l. The product of these three quantities is to be multiplied by a certain proportionality factor, which still depends on the four quantities k, l, κ, λ that determine the nature of the collision, but not on the time; we denote it by $A_{\kappa\lambda}^{kl}$. Thus we write

$$N_{\kappa\lambda}^{kl} = \tau w_k w_l A_{\kappa\lambda}^{kl} \tag{28}$$

The number of collisions is analogous to what we determined earlier in equation (2). The quantity A appears here in place of what was earlier denoted by ψ. If we wish to make the analogy complete, we must attribute to A the properties of ψ. ψ satisfies the equation

$$\sqrt{xx'}\psi(x, x', \xi) = \sqrt{\xi(x+x'-\xi)}\psi(\xi, x+x'-\xi, x) \tag{29}$$

In our case the kinetic energies before the collision are $k\varepsilon$, $l\varepsilon$, and after the collision $\kappa\varepsilon$, $\lambda\varepsilon$; in our case, therefore,

$$x = k\varepsilon; \quad x' = l\varepsilon; \quad \xi = \kappa\varepsilon; \quad x+x'-\xi = \lambda\varepsilon$$

The quantity $\psi(x, x', \xi)$ corresponds to $A_{\kappa\lambda}^{kl}$, and one sees easily that the quantity $\psi(\xi, x+x'-\xi, x)$ corresponds to $A_{kl}^{\kappa\lambda}$. Equation (29) therefore reduces in our case to

$$\sqrt{kl}A_{\kappa\lambda}^{kl} = \sqrt{\kappa\lambda}A_{kl}^{\kappa\lambda} \tag{30}$$

Now the analogy is complete, and we need only to make ε infinitely small and $p\varepsilon$ infinitely large in order to obtain from the solution of this problem the solution of the physical one treated earlier. The formulae will be somewhat simpler if we denote $\sqrt{kl}A_{\kappa\lambda}^{kl}$ by $B_{\kappa\lambda}^{kl}$. Then equation (30) reduces to

$$B_{\kappa\lambda}^{kl} = B_{kl}^{\kappa\lambda} \tag{31}$$

and equation (28) becomes

$$N_{\kappa\lambda}^{kl} = \tau \frac{w_k w_l}{\sqrt{kl}} B_{\kappa\lambda}^{kl} \tag{32}$$

The square roots are of course to be taken positive, since $N_{\kappa\lambda}^{kl}$ and the w's are essentially positive numbers, and we wish to choose the B's always positive.

After these preparations, we ask what change the quantity w_1 experiences during the time τ. w_1 is the number of molecules with kinetic energy ε in unit volume. We know that this number changes only through collisions. Every time two molecules collide such that before the collision one of them has kinetic energy ε, while afterwards it no longer has that energy, this number decreases by one. Conversely, every time two molecules collide in such a way that before the collision neither, but after it one of them has kinetic energy ε, this number increases by one. If we subtract the former number from w_1 and add the latter to it, we obtain the number of molecules in unit volume which have kinetic energy ε at time $t+\tau$. We now have to find out how many collisions there are in which one molecule has kinetic energy ε before the collision. When the other one also has kinetic energy ε, then after the collision each must still have kinetic energy ε, since the total kinetic energy of the two must still be 2ε, and no other kinetic energies than those listed in (26) can occur. If one molecule has kinetic energy ε before the collision and the other has kinetic energy 2ε, then for the same reason, after the collision one must have ε and the other 2ε. None of these collisions change the number of molecules with energy ε. However, if one has ε and the other has 3ε before the collision, then after the collision both may have 2ε. Each such collision will decrease w_1 by one. There will be N_{22}^{13} of these collisions in unit volume during time τ, so that w_1 decreases by N_{22}^{13}. We have to subtract N_{22}^{13} from w_1. Likewise we have to subtract N_{23}^{14}, N_{32}^{14}, N_{24}^{15}, ..., $N_{p-1,2}^{1p}$ from w_1. On the other hand we have to add the numbers N_{13}^{22}, N_{14}^{23}, ..., $N_{1p}^{p-1,2}$, since through each of these collision the number of molecules with kinetic energy ε increases by one. Thus we obtain

$$
\begin{aligned}
w_1' = w_1 &- N_{22}^{13} - N_{23}^{14} - N_{32}^{14} - N_{24}^{15} - \ldots, \\
&+ N_{13}^{22} + N_{14}^{23} + N_{14}^{32} + N_{15}^{24} + \ldots
\end{aligned} \tag{33}
$$

The law of formation of the series is easily grasped. We have to subtract all the N's which have an upper index 1, and add all those that have a lower index 1. Those which have this index both above and below are to be added and subtracted both, so that they may be left out entirely. (Earlier in the integral we have not eliminated these self-cancelling terms, for the sake of convenience.) We must also observe that the four indices of N must satisfy equation (27), and that two N's which are the same if one permutes the upper and lower indices simultaneously (e.g. N_{23}^{14} and N_{32}^{41}) correspond to completely identical collisions, and hence should be added (or subtracted) only once.

If we expand w_1' according to Taylor's theorem, we obtain

$$w_1' = w_1 + \tau \frac{dw_1}{dt}$$

If we substitute this as well as the value of N given by equation (32) into equation (33), then we obtain, after dividing through by τ:

$$\frac{dw_1}{dt} = -B_{22}^{13}\frac{w_1 w_3}{\sqrt{1}\sqrt{3}} - B_{23}^{14}\frac{w_1 w_4}{\sqrt{1}\sqrt{4}} - B_{32}^{14}\frac{w_1 w_4}{\sqrt{1}\sqrt{4}} - B_{24}^{15}\frac{w_1 w_5}{\sqrt{1}\sqrt{5}} - \cdots$$

$$+ B_{13}^{22}\frac{(w_2)^2}{2} + B_{14}^{23}\frac{w_2 w_3}{\sqrt{2}\sqrt{3}} + B_{14}^{32}\frac{w_2 w_3}{\sqrt{2}\sqrt{3}} + B_{15}^{24}\frac{w_2 w_4}{\sqrt{2}\sqrt{4}} + \cdots$$

which can also be written, taking account of equation (32), as:

$$\frac{dw_1}{dt} = B_{22}^{13}\left(\frac{(w_2)^2}{2} - \frac{w_1 w_3}{\sqrt{1}\sqrt{3}}\right) + (B_{23}^{14} + B_{32}^{14})\left(\frac{w_2 w_3}{\sqrt{2}\sqrt{3}} - \frac{w_1 w_4}{\sqrt{1}\sqrt{4}}\right) + \cdots$$

Likewise one finds

$$\frac{dw_2}{dt} = 2B_{22}^{13}\left(\frac{w_1 w_3}{\sqrt{1}\sqrt{3}} - \frac{(w_2)^2}{2}\right)$$

$$+ (B_{23}^{14} + B_{32}^{14})\left(\frac{w_1 w_4}{\sqrt{1}\sqrt{4}} - \frac{w_2 w_3}{\sqrt{2}\sqrt{3}}\right) + \cdots$$

$$\frac{dw_p}{dt} = (B_{2,\,p-1}^{1,\,p} + B_{p-1,\,2}^{1,\,p})\left(\frac{w_{p-1} w_2}{\sqrt{p-1}\sqrt{2}} - \frac{w_1 w_p}{\sqrt{1}\sqrt{p}}\right)$$

$$+ (B_{3,\,p-2}^{1,\,p} + B_{p-2,\,3}^{1,\,p})\left(\frac{w_3 w_{p-2}}{\sqrt{3}\sqrt{p-2}} - \frac{w_1 w_p}{\sqrt{1}\sqrt{p}}\right) + \cdots$$

$$(34)$$

All that needs to be explained is why the term

$$B_{22}^{13} \frac{w_1 w_3}{\sqrt{1}\sqrt{3}}$$

has the factor 2 in the expression for dw_2/dt. This term comes from collisions for which the kinetic energies are ε, 3ε before the collision and 2ε, 2ε after the collision; each such collision changes the number of molecules with kinetic energy 2ε by two rather than by one, since two molecules simultaneously acquire kinetic energy 2ε. Hence, all these collisions must be counted twice. Likewise in the expression for dw_3/dt, the terms

$$B_{33}^{15} \frac{w_1 w_5}{\sqrt{1}\sqrt{5}} \quad \text{and} \quad B_{33}^{24} \frac{w_2 w_4}{\sqrt{2}\sqrt{4}}$$

are counted twice, and so forth.

It would be easy to represent the system of equations (34) by summation formulae; however, I believe that this would not make it any easier to understand; the law of formation of the terms is already clear enough. One also sees that this is exactly the system of equations to which the partial differential equation (18) is transformed when one uses the Lagrange method, as previously explained, to replace it by a system of p ordinary differential equations, and denotes $f(k\varepsilon, t)$ by w_k. In order to simplify equations (34) somewhat, we set

$$w_k = \sqrt{k} u_k$$

These equations then become:

$$
\left.
\begin{aligned}
\frac{du_1}{dt} &= B_{22}^{13}(u_2^2 - u_1 u_3) + (B_{23}^{14} + B_{32}^{14})(u_2 u_3 - u_1 u_4) + \cdots \\[2mm]
\sqrt{2}\frac{du_2}{dt} &= 2B_{22}^{13}(u_1 u_3 - u_2^2) + (B_{23}^{14} + B_{32}^{14})(u_1 u_4 - u_2 u_3) + \cdots \\
&\qquad\qquad \vdots \\
\sqrt{p}\frac{du_p}{dt} &= (B_{2,\,p-1}^{1,\,p} + B_{p-1,\,2}^{1,\,p})(u_2 u_{p-1} - u_1 u_p) + \cdots
\end{aligned}
\right\} \quad (35)
$$

From these equations it can again be proved that

$$E = u_1 \log u_1 + \sqrt{2} u_2 \log u_2 + \ldots + \sqrt{p} u_p \log u_p$$

must always decrease unless $u_2^2 - u_1 u_3$, $u_2 u_3 - u_1 u_4$, ... (in other words all the expressions multiplied by the coefficients B in equation (35)) vanish. Equations (35) have the inconvenient feature that while they can be written with summation formulae they cannot be written out completely explicitly. It would undoubtedly be an aid to clarity, therefore, if we begin with the simplest case and then proceed gradually to the general case. First let $p = 3$; the molecules can only have three different kinetic energies, ε, 2ε, and 3ε. Then the system of equations (35) reduces to the following three equations:

$$\left.\begin{array}{l}
\dfrac{du_1}{dt} = B_{22}^{13}(u_2^2 - u_1 u_3) \\[2em]
\sqrt{2}\,\dfrac{du_2}{dt} = 2B_{22}^{13}(u_1 u_3 - u_2^2) \\[2em]
\sqrt{3}\,\dfrac{du_3}{dt} = B_{22}^{12}(u_2^2 - u_1 u_3)
\end{array}\right\} \quad (36)$$

and the expression for E reduces to

$$E = u_1 \log u_1 + \sqrt{2} u_2 \log u_2 + \sqrt{3} u_3 \log u_3$$

The differentiation gives

$$\frac{dE}{dt} = (\log u_1 + 1)\frac{du_1}{dt} + \sqrt{2}(\log u_2 + 1)\frac{du_2}{dt} + \sqrt{3}(\log u_3 + 1)\frac{du_3}{dt}$$

E*

or, with a different arrangement of the terms,

$$\frac{dE}{dt} = \log u_1 \frac{du_1}{dt} + \sqrt{2} \log u_2 \frac{du_2}{dt} + \sqrt{3} \log u_3 \frac{du_3}{dt}$$

$$+ \frac{du_1}{dt} + \sqrt{2}\frac{du_2}{dt} + \sqrt{3}\frac{du_3}{dt}$$

The sum of the last three terms vanishes according to equations (36) so that one obtains dE/dt by multiplying the first of these equations by $\log u_1$, the second by $\log u_2$, and the third by $\log u_3$, and adding all three together. If one does this he obtains

$$\frac{dE}{dt} = B_{22}^{13}.(u_2^2 - u_1 u_3).(\log u_1 + \log u_3 - 2\log u_2)$$

or

$$\frac{dE}{dt} = B_{22}^{13}.(u_2^2 - u_1 u_3) \log \left(\frac{u_1 u_3}{u_2^2}\right)$$

Of the two factors multiplying B_{22}^{13} on the right-hand side of this equation, the first is positive and the second is negative when $u_2^2 > u_1 u_2$, whereas the first is negative and the second is positive when $u_2^2 > u_1 u_2$. Hence their product is always negative, and since B_{22}^{13} must be positive, dE/dt is always negative or zero. The latter is true when $u_2^2 = u_1 u_3$. Now it can easily be shown that E cannot become negatively infinite. Obviously none of the three quantities u_1, u_2, u_3 may be negative or imaginary. For positive u, however, $u \log u$ cannot have a larger negative value than $-1/e$, hence E cannot have a larger negative value than

$$-\frac{1 + \sqrt{2} + \sqrt{3}}{e}$$

where e is the base of natural logarithms.

Therefore E, since its derivative cannot be positive, must continually approach a minimum for which $dE/dt = 0$, and for which $u_2^2 = u_1 u_3$. The proof cannot be carried out in just the same way when $n > 3$. I consider here only the case $n = 4$. In this case equations (35) reduce to

$$\frac{du_1}{dt} = B_{22}^{13}(u_2^2 - u_1 u_3) + (B_{23}^{14} + B_{32}^{14})(u_2 u_3 - u_1 u_4)$$

$$\sqrt{2}\,\frac{du_2}{dt} = 2B_{22}^{13}(u_1 u_3 - u_2^2) + (B_{23}^{14} + B_{32}^{14})(u_1 u_4 - u_2 u_3)$$
$$+ B_{33}^{24}(u_3^2 - u_2 u_4)$$

$$\sqrt{3}\,\frac{du_3}{dt} = B_{22}^{13}(u_2^2 - u_1 u_3) + (B_{23}^{14} + B_{32}^{14})(u_1 u_4 - u_2 u_3)$$
$$+ 2B_{33}^{24}(u_2 u_4 - u_3^2)$$

$$\sqrt{4}\,\frac{du_4}{dt} = (B_{23}^{14} + B_{32}^{14})(u_2 u_3 - u_1 u_4) + B_{33}^{24}(u_3^2 - u_2 u_4)$$

$$\left.\right\} \quad (37)$$

For E one finds

$$E = u_1 \log u_1 + \sqrt{2} u_2 \log u_2 + \sqrt{3} u_3 \log u_3 + \sqrt{4} u_4 \log u_4$$

$$\frac{dE}{dt} = \log u_1 \frac{du_1}{dt} + \sqrt{2} \log u_2 \frac{du_2}{dt} + \sqrt{3} \log u_3 \frac{du_3}{dt}$$
$$+ \sqrt{4} \log u_4 \frac{du_4}{dt}$$

If one substitutes here for

$$\frac{du_1}{dt}, \quad \frac{du_2}{dt}, \quad \frac{du_3}{dt}, \quad \frac{du_4}{dt}$$

their values from equations (37), he obtains, with a suitable rearrangement of terms,

$$\frac{dE}{dt} = B_{22}^{13}(u_2^2 - u_1 u_3) \log\left(\frac{u_1 u_3}{u_2^2}\right) + B_{33}^{24}(u_3^2 - u_2 u_4) \log\left(\frac{u_2 u_4}{u_3^2}\right)$$
$$+ (B_{23}^{14} + B_{32}^{14})(u_2 u_3 - u_1 u_4) \log\left(\frac{u_1 u_4}{u_2 u_3}\right)$$

I remark that the change in the order of the summands, which is necessary here, is analogous to our previous transformation of definite integrals. From the above expression one sees at once that dE/dt is again necessarily negative, unless simultaneously we have

$$u_2^2 = u_1 u_3, \ u_3^2 = u_2 u_4, \ u_2 u_3 = u_1 u_4$$

which can also be written

$$u_3 = \frac{u_2^2}{u_1}, \quad u_4 = \frac{u_2^3}{u_1^2}$$

Likewise one finds in the general case that dE/dt is necessarily negative so that E must decrease unless

$$u_3 = \frac{u_2^2}{u_1}, \quad u_4 = \frac{u_2^3}{u_1^2}, \dots \tag{38}$$

Since E cannot have a larger negative value than

$$-\frac{1 + \sqrt{2} + \sqrt{3} + \dots + \sqrt{p}}{e} \tag{39}$$

it must necessarily approach a minimum value for which equations (38) hold. Thus it continually approaches the distribution of states determined by equations (38). We now have to prove that equations (38) uniquely determine the distribution of states. If we add together all the equations (35), we obtain

$$\frac{du_1}{dt} + \sqrt{2}\frac{du_2}{dt} + \sqrt{3}\frac{du_3}{dt} + \dots + \sqrt{p}\frac{du_p}{dt} = 0$$

hence

$$u_1 + \sqrt{2}u_2 + \sqrt{3}u_3 + \dots + \sqrt{p}u_p = a \tag{40}$$

In a similar way we find that

$$u_1 + 2\sqrt{2}u_2 + 3\sqrt{3}u_3 + \dots + p\sqrt{p}u_p = \frac{b}{\varepsilon} \tag{41}$$

where a and b are constants. The meaning of these equations is obvious. In particular,

$$w_1 + w_2 + w_3 + \ldots = u_1 + \sqrt{2}u_3 + \sqrt{3}u_3 + \ldots = a$$

is the total number of molecules in unit volume, while b is their total kinetic energy. Equations (40) and (41) therefore tell us that these two quantities are constant. Suppose that the two quantities a and b are given. Then we set the quotient u_2/u_1 equal to γ. Equations (38) then reduce to

$$u_3 = \gamma^2 u_1, \ u_4 = \gamma^3 u_1, \ \ldots, \ u_p = \gamma^{p-1} u_1$$

If one substitutes these values into equations (40) and (41), then he finds easily

$$\left. \begin{array}{l} \left(pa - \dfrac{b}{\varepsilon}\right)\sqrt{p}\gamma^{p-1} + \left[(p-1)a - \dfrac{b}{\varepsilon}\right]\sqrt{p-1}\gamma^{p-2} + \ldots \\[4mm] + \left(3a - \dfrac{b}{\varepsilon}\right)\sqrt{3}\gamma^2 + \left(2a - \dfrac{b}{\varepsilon}\right)\sqrt{2}\gamma + a - \dfrac{b}{\varepsilon} = 0 \end{array} \right\} \quad (42)$$

Since all the u's are necessarily positive, we see immediately that $(b/\varepsilon) - a$ must be positive while $(b/\varepsilon) - pa$ must be negative. Hence b must lie between εa and εpa. Hence, in equation (42) the coefficient of γ^{p-1} is positive, while the term independent of γ must be negative. The polynomial is therefore positive for $\gamma = \infty$, and negative for $\gamma = 0$; therefore there is one and only one positive root for γ, since the series of coefficients changes sign only once. Negative or imaginary values for γ are of course meaningless. But from γ we can determine uniquely all the u's and also all the w's. Hence, whatever may be the initial distribution of states, there is one and only one distribution which it approaches with increasing time. This distribution depends only on the constants a and b, the total number and total kinetic energy of the molecules (density and temperature of the gas). This theorem was proved first only for the case that the distribution of states is initially uniform. It must also hold, however, when this is not true, provided only that the molecules are distributed in such

130   KINETIC THEORY

a way that they tend to become mixed as time progresses, so that
the distribution becomes uniform after a very long time. This will
always happen with the exception of certain special cases, for
example, when the molecules move initially in a straight line and
are reflected back in this straight line at the walls. Since we have
established this for arbitrary p and ε, we can immediately go to the
case where $1/p$ and ε become infinitesimal.† We have first:

$$w_k = \sqrt{k}u_k = u_1\sqrt{k}\gamma^{k-1}$$

† For very large p, the expression (39) will be very large, of order $p^{\frac{3}{2}}$. In
this case it is necessary to look for a smaller negative value that E can never
exceed. The quantity denoted here by E differs by a constant from the one
earlier so denoted. If we wish to obtain the quantity denoted by E_1 in
equation (17a), page 113, which again differs only by a constant from the other
quantities denoted by this letter, then we must add to our present E,

$$-\frac{3 \log \varepsilon}{2}(u_1 + \sqrt{2}u_2 + \ldots)$$

Therefore

$$E_1 = E - \frac{3 \log \varepsilon}{2}(u_1 + \sqrt{2}\,u_2 + \ldots) = u_1 \log \frac{u_1}{\varepsilon^{\frac{3}{2}}} + \sqrt{2}\,u_2 \log \frac{u_2}{\varepsilon^{\frac{3}{2}}} + \ldots$$

It is clear now that E_1 is a real and continuous function of the u's for all real
positive values of u. Furthermore, if we say that a negative quantity is smaller,
the greater its numerical value is, then E is not smaller than the expression
(39), hence E_1 is not smaller than

$$-\frac{1}{2}(1 + \sqrt{2} + \ldots \sqrt{p}) - \frac{3}{2}a \log \varepsilon$$

Hence, E_1 must have a minimum if the u's run through all real positive values
compatible with equations (40) and (41). One can then easily show that for
this minimum none of the u's can be equal to zero, so that the minimum cannot
lie on the boundary of the space formed from the u's, and consequently it can
be found by applying the usual rules of differential calculus. If we add to the
total differential of E_1 that of the two equations (40) and (41), multiplying the
former with the undetermined multiplier λ, and the latter by μ, then we obtain

$$(\log u_1 + \lambda + \mu)\,du_1 + (\log u_2 + \lambda + 2\mu)\,\sqrt{2}\,du_2 + \ldots = 0$$

At the minimum, the factor of each differential must vanish, whence on
elimination of λ and μ one obtains

$$\log u_2 - \log u_1 = \log u_3 - \log u_2 = \ldots$$

or

$$u_3 = \frac{u_2^2}{u_1}, \quad u_4 = \frac{u_3^2}{u_2} \ldots$$

(Footnote continued on next page

For infinitesimal ε we can again set

$$\varepsilon = dx, \ k\varepsilon = x, \ \gamma = e^{-h\varepsilon}, \ \frac{u_1}{\gamma \varepsilon^{\frac{3}{2}}} = C \tag{43}$$

and obtain

$$w_k = C\sqrt{x}e^{-hx}dx$$

which is again the Maxwell distribution. Likewise one can convince himself that the sum which we have here denoted by E

---

which we recognize to be the same as equations (38). These equations therefore determine the smallest value that E_1 can have when the u's take all possible values consistent with equations (40) and (41). However, since the u's are actually subject to equations (40 and 41) during the entire process, this is the smallest value of E_1 during the entire process. In order to calculate it, we set again

$$u_2 = u_1\gamma, u_3 = u_1\gamma^2 \dots$$

We know that we then find from equations (38), (40) and (41) a unique positive value for γ, which must correspond to the actual minimum of E_1. This minimum value of E_1 is therefore

$$E = \frac{1}{\varepsilon} b \log \gamma + a \log \left(\frac{u_1}{\gamma \varepsilon^{\frac{3}{2}}}\right)$$

E_1 cannot have a smaller value than this. This value remains finite for infinitesimal ε and infinite p. Taking account of equations (43), we see that it reduces to

$$a \log C - bh$$

or, since

$$a = \frac{1}{2}\sqrt{\frac{\pi}{h^3}} \ C, b = \frac{3a}{2h}$$

one can write for it,

$$\frac{1}{2}\sqrt{\frac{\pi}{h^3}} \ C\left(\log C - \frac{3}{2}\right)$$

which, since the constants C and h are not infinite, is a finite quantity. Hence E_1 cannot be minus infinity. On the other hand, it may be plus infinity. We still have to show that in that case there cannot be thermal equilibrium. This proof, as well as an explicit discussion of the exceptional case where

$$\lim (\varepsilon/\tau)[f(\varepsilon, t+\tau) \log f(\varepsilon, t+\tau) + \sqrt{2} f(2\varepsilon, t+\tau) \log f(2\varepsilon, t+\tau) + \dots$$
$$-f(\varepsilon, t) \log f(\varepsilon, t) - \sqrt{2} f(2\varepsilon, t) \log f(\varepsilon, t) - \dots]$$

comes out to be different according as ϵ/τ or τ/ϵ vanishes, will not be discussed further here.

reduces, aside from a constant additive term, to the integral in equation (17a); we therefore obtain by this method all the results that we earlier found by transformations of definite integrals, but it is the advantage of being much simpler and clearer. One only has to accept the abstraction that a molecule may have only a finite number of kinetic energies as a transition stage.

If one sets the time derivatives in equations (35) equal to zero, he obtains the conditions that the distribution of states does not change with time but is stationary. One sees also that equations (35) have many other solutions in addition to the one we have found, but these do not represent acceptable stationary distributions since the probabilities of certain kinetic energies comes out to be negative or imaginary. The same is true when, as actually happens in nature, each molecule can have any kinetic energy from zero to infinity. The condition that the distribution be stationary is obtained by setting

$$\frac{\partial f(x,t)}{\partial t} = 0$$

in equation (16). This gives

$$0 = \int\limits_{0}^{\infty} \int\limits_{0}^{x+x'} \left[\frac{f(\xi)f(x+x'-\xi)}{\sqrt{\xi}\sqrt{x+x'-\xi}} - \frac{f(x)f(x')}{\sqrt{xx'}} \right] \sqrt{xx'} \psi(x,x',\xi) dx' d\xi$$

A solution of this equation is

$$f(x) = C\sqrt{x}e^{-hx}$$

which is the Maxwell distribution. From what has been said previously it follows that there are infinitely many other solutions, which are not useful however since $f(x)$ comes out negative or imaginary for some values of x. Hence, it follows very clearly that Maxwell's attempt to prove *a priori* that his solution is the only one must fail, since it is not the only one but rather it is the only one that gives purely positive probabilities, and therefore it is the only useful one.

III. Diffusion, viscosity, and heat conduction of a Gas

Here we shall make room for a few remarks pertaining to the case that the distribution is not completely irregular, but is still not what we have called uniform, so that not all directions of velocity are equivalent; this corresponds to the case of viscosity and heat conduction. Let

$$f(\xi, \eta, \zeta, x, y, z, t)d\xi d\eta d\zeta$$

be the number of molecules in unit volume at the position (x, y, z) in the gas for which the velocity components in the x-direction lie between ξ and $\xi + d\xi$, those in the y-direction lie between η and $\eta + d\eta$, and those in the z-direction lie between ζ and $\zeta + d\zeta$. A collision is determined by the velocity components ξ, η, ζ and ξ_1, η_1, ζ_1 of the two colliding molecules before the collision and by the quantities b and ϕ. (The latter two quantities, as well as V, k, A_2, X, etc., which will appear later, have the same meaning as in Maxwell's paper.†) The velocity components after the collision: ξ, η, ζ and ξ_1, η_1, ζ_1 are functions of these eight variables. If we write for brevity $d\omega_1$ for $d\xi_1 \, d\eta_1 \, d\zeta_1$ and denote by f the value of the function $f(\xi, \eta, \zeta, x, y, z, t)$, and by f_1, f' and f_1' the values of this function when one substitutes for (ξ, η, ζ) the variables (ξ_1, η_1, ζ_1), (ξ', η', ζ') or $(\xi_1', \eta_1', \zeta_1')$ respectively, then the function f must satisfy the differential equation

$$\left. \begin{array}{l} \dfrac{\partial f}{\partial t} + \xi \dfrac{\partial f}{\partial x} + \eta \dfrac{\partial f}{\partial y} + \zeta \dfrac{\partial f}{\partial z} + X \dfrac{\partial f}{\partial \xi} + Y \dfrac{\partial f}{\partial \eta} + Z \dfrac{\partial f}{\partial \zeta} \\[2mm] + \int d\omega_1 \int b \, db \int d\phi \, V(ff_1 - f'f_1') = 0 \end{array} \right\} \quad (44)$$

This can easily be seen by imagining that the volume element moves with velocity (ξ, η, ζ) and considering how the distribution of states is changed by collisions. If the gas is enclosed by fixed walls, then it follows again from equation (44) that E can only be decreased by molecular motion, if one sets

$$E = \iiint\!\!\iiint f \log f \, dx \, dy \, dz \, d\xi \, d\eta \, d\zeta$$

---

† J. C. Maxwell, *Phil. Mag.* **35**, 129, 185 (1868) [Selection 1].

which expression is proportional to the entropy of the gas. In order to give just one example for the case of different boundary conditions, let the repulsion of two molecules be inversely proportional to the fifth power of their distance. X, Y, and Z will always be zero in the following. We shall set

$$f = A(1 + 2ha y\xi + c\xi\eta)e^{-h(\xi^2 + \eta^2 + \zeta^2)} \tag{45}$$

where the two constants a and c are very small. If we substitute this value into equation (44), neglect squares and products of a and c, and carry out the integration over b and ϕ just as Maxwell has done,† then we find that equation (44) is satisfied when

$$c = -\frac{2ha}{3A_2 k\rho}$$

The formula (45) thus gives a possible distribution of states: one in which each layer parallel to the xz plane moves in the direction of the x axis with velocity ay, if y is the y-coordinate of that layer. This is the simplest case of viscous flow. The viscosity constant is then the momentum transported through unit surface in unit time divided by $-a$:

$$-\frac{\rho\bar{\xi\eta}}{a} = -\frac{\rho}{a}\frac{\iiint \xi\eta f \, d\xi d\eta d\zeta}{\iiint f \, d\xi d\eta d\zeta} = \frac{1}{6A_2 kh} = \frac{p}{3A_2 k\rho}$$

just as Maxwell has already found. The notations are those of Maxwell. A more general expression is the following:

$$f = A\left[1 - \frac{2ht}{3}\left(\frac{\partial u}{\partial x} + \frac{\partial v}{\partial y} \quad \frac{\partial w}{\partial z}\right)(\xi^2 + \eta^2 + \zeta^2)\right.$$
$$+ 2h(u\xi + v\eta + w\zeta) + \alpha\xi^2 + \beta\eta^2 + \gamma\zeta^2 + \alpha'\eta\zeta + \beta'\xi\zeta$$
$$\left. + \gamma'\xi\eta \right]e^{-h(\xi^2 + \eta^2 + \zeta^2)} \tag{46}$$

---

† Maxwell, *op. cit.* p. 141-4.

This expression likewise satisfies equation (44), when u, v, and w are linear functions of x, y, z, and

$$\alpha = -\frac{2h}{3A_2k\rho}\frac{\partial u}{\partial x}, \quad \alpha' = -\frac{2h}{3A_2k\rho}\left(\frac{\partial v}{\partial z}+\frac{\partial w}{\partial z}\right)$$

β, γ, β' and γ' have similar values. The expression (46) represents an arbitrary motion of the gas in which the velocity components u, v, w at the point with coordinates x, y, z are linear functions of these coordinates. Unless

$$\frac{\partial u}{\partial x}+\frac{\partial v}{\partial y}+\frac{\partial w}{\partial z} = 0$$

the density and temperature change with time.

If one calculates

$$\overline{\xi^2}, \ \overline{\eta^2}, \ \overline{\xi\eta} \ldots$$

using the expression (46), he obtains again the values found by Maxwell. If $\partial^2 u/\partial x^2$ is different from zero, one obtains one more term in equation (44) which does not vanish, namely

$$\frac{\partial\alpha}{\partial x}\cdot\xi^3 e^{-h(\xi^2+\eta^2+\zeta^2)}$$

whose average value we can take to be approximately

$$\left.\begin{aligned}
-\frac{\partial\alpha}{\partial x}(\overline{\xi^2})^{\frac{3}{2}}e^{-h(\xi^2+\eta^2+\zeta^2)} &= \frac{2h}{3A_2k\rho}\sqrt{\frac{p^3}{\rho^3}}\,e^{-h(\xi^2+\eta^2+\zeta^2)}\frac{\partial^2 u}{\partial x^2} \\
&= \frac{\mu}{\rho}\sqrt{\frac{\rho}{p}\frac{\partial^2 u}{\partial x^2}}\,e^{-h(\xi^2+\eta^2+\zeta^2)}
\end{aligned}\right\} \quad (47)$$

One sees easily that this term vanishes in comparison to the other terms in equation (44) if one substitutes therein the value of f from equation (46), so that equation (44) is still approximately satisfied. In calculating the distribution of states one should expand the quantities u, v, w in a Taylor series and retain the first powers of

x, y, and z. The first term of (44), after substituting the value of f from (46), would thus be

$$-\frac{2h}{3}\xi^2\frac{\partial u}{\partial x}e^{-h(\xi^2+\eta^2+\zeta^2)}$$

Its average value is therefore

$$\frac{1}{3}\frac{\partial u}{\partial x}e^{-h(\xi^2+\eta^2+\zeta^2)}$$

If one calculates the ratio of (47) to this quantity, for air at 0°C and normal atmospheric pressure, he finds that it is about

$$0\cdot00009 \text{ mm} \frac{\partial^2 u/\partial x^2}{\partial u/\partial x}$$

It is therefore negligible when

$$\frac{\partial u/\partial x}{\partial^2 u/\partial x^2}$$

is about 1 mm, in other words if values of $\partial u/\partial x$ at points 1 mm apart are in the ratio 1 : 2 on the average. Only when $\partial u/\partial x$ starts to change this much over distances of the order of a mean free path will the ratio become significant.

The value

$$f = A[1+ax+by+cz-(a\xi+b\eta+c\zeta)t]\,e^{-h(\xi^2+\eta^2+\zeta^2)}$$

also satisfies equation (44).

For a mixture of two kinds of gas, we shall indicate quantities pertaining to the second kind by an asterisk, so that p and p_* will be the partial pressures, m and m_* the masses of molecules of the two kinds. Then in place of equation (44) we have:

$$\left.\begin{aligned}&\frac{\partial f}{\partial t}+\frac{\partial f}{\partial x}+\frac{\partial f}{\partial y}+\frac{\partial f}{\partial z}+\int d\omega_1\int b\,db\int d\phi V(ff_1-f'f'_1)\\&\qquad+\int d\omega_*\int b\,db\int d\phi V(ff_*-f'f'_*)=0\end{aligned}\right\} \quad (44^*)$$

and a similar equation for the second kind of gas. The simplest case of diffusion corresponds to

$$f = \sqrt{m^3 h^3/\pi^3}\, N(1+2hmu\xi)\, e^{-hm(\xi^2+\eta^2+\zeta^2)}$$
$$f_* = \sqrt{m_*^3 h^3/\pi^3}\, N_*(1+2hm_*u\xi)\, e^{-hm_*(\xi^2+\eta^2+\zeta^2)}$$

$$(46^*)$$

where N and N_* are functions of x, but Nu and N_*u_* are constant. None of these quantities depends on time. Equation (44*) is satisfied when

$$\frac{dN}{dx} + NN_* 2hmm_*(u-u_*)A_1 k = 0$$

A similar equation must hold for N_*. We must also have $N+N_* = $ constant = number of molecules of both kinds in unit volume; hence, it follows that $Nu = - N_*u_* = $ number of molecules of one kind of gas that go through a unit cross-section in unit time. The diffusion constant is

$$-\frac{Nu}{dN/dx} = \frac{1}{(N+N_*)2hmm_*A_1 k} = \frac{pp_*}{A_1 k\rho\rho_*(p+p_*)}$$

since

$$2h = \frac{N+N_*}{p+p_*} = \frac{N}{p} = \frac{N_*}{p_*}$$

To obtain the equations of motion, one multiplies equation (44) or (44*) by $m\xi d\omega$ (where $d\omega = d\xi\, d\eta\, d\zeta$) and integrates over all ξ, η, and ζ. The first four terms of these equations become

$$\frac{\partial(\rho u)}{\partial t} + \frac{\partial(\rho\overline{\xi^2})}{\partial x} + \frac{\partial(\rho\overline{\xi\eta})}{\partial y} + \frac{\partial(\rho\overline{\xi\zeta})}{\partial z}$$

or since

$$\frac{\partial\rho}{\partial t} + \frac{\partial(\rho u)}{\partial x} + \frac{\partial(\rho v)}{\partial y} + \frac{\partial(\rho w)}{\partial z} = 0$$

they become

$$\rho\left(\frac{\partial u}{\partial t} + u\frac{\partial u}{\partial x} + v\frac{\partial u}{\partial y} + w\frac{\partial u}{\partial z}\right) + \frac{\partial(\rho\overline{\xi'^2})}{\partial x} + \frac{\partial(\rho\overline{\xi'\eta'})}{\partial y} + \frac{\partial(\rho\overline{\xi'\zeta'})}{\partial z}$$

where we have set

$$\xi = \xi' + u, \ \eta = \eta' + v, \ \zeta = \zeta' + w$$

The other terms are just minus the momentum transported by molecular collisions, which is zero if no second gas is present. The momentum transport, aside from that resulting from the pressure forces,

$$\left(-\frac{\partial(\rho\overline{\xi'^2})}{\partial x} - \frac{\partial(\rho\overline{\xi'\eta'})}{\partial y} - \frac{\partial(\rho\overline{\xi'\zeta'})}{\partial z} \right)$$

is therefore equal to the acceleration multiplied by the density:

$$\left(\frac{\partial u}{\partial t} + u\frac{\partial u}{\partial x} + v\frac{\partial u}{\partial y} + w\frac{\partial u}{\partial z} \right)$$

The latter equations are valid for any force law. On the other hand, the expressions (45), (46) and (46*) are correct only when the repulsion between two molecules is inversely proportional to the fifth power of their distance. For any other force law—for example, when the molecules bounce off each other like elastic spheres—the expressions (45), (46) and (46*) do not satisfy equations (44) and (44*), so that for other force laws the velocity distribution for diffusion, viscosity, etc., does not have such a simple form. For the case of diffusion, we have to represent f in the following form:

$$A[1 + a\xi + b\xi^3 + c(\eta^2 + \zeta^2)\xi + d\xi^5 \dots]e^{-h(\xi^2 + \eta^2 + \zeta^2)} \qquad (47^*)$$

and I see no other way to solve equation (44*) than by successive determination of the coefficients a, b, c ... For all other force laws, therefore, the velocity distribution of a diffusing gas is not the same as if it were moving alone in space with its diffusion velocity u. This is because molecules with different velocities also have different diffusion velocities, so that the velocity distribution will be continually disturbed. Since the terms in the expression (47*) with ξ^3, $\xi\eta^2$, ..., lead to terms in the diffusion constant that are of the same order of magnitude as the one with ξ, the diffusion constant cannot be obtained accurately by first leaving out the terms for momentum transport. Nevertheless, the error thereby incurred should scarcely be very large. The same holds for viscosity and heat conduction. Indeed it is not merely the values of the diffusion, viscosity, etc., constants that are in question here, but rather their constancy, in the case of force laws other than Maxwell's.

The case of heat conduction in the direction of the x-axis corresponds to the following value of f, assuming Maxwell's force law:

$$f = A[1 + ax(\xi^2 + \eta^2 + \zeta^2) + bx + c\xi + g\xi(\xi^2 + \eta^2 + \zeta^2)]e^{-h(\xi^2 + \eta^2 + \zeta^2)}$$

whence it follows that

$$\xi\frac{\partial f}{\partial x} + \eta\frac{\partial f}{\partial y} + \zeta\frac{\partial f}{\partial z} = \xi A e^{-h(\xi^2 + \eta^2 + \zeta^2)}[a(\xi^2 + \eta^2 + \zeta^2) + b]$$

When one substitutes this value of f into equation (44) and carries out all the integrations according to Maxwell's procedure, the last term of (44) reduces to

$$2gA_2kMN\xi\left(\xi^2 + \eta^2 + \zeta^2 - \frac{5}{2h}\right)Ae^{-h(\xi^2 + \eta^2 + \zeta^2)}$$

In order that equation (44) may be satisfied, we must take

$$a = -2gA_2kMN, \quad b = 5gA_2kMN.\frac{1}{h} = -\frac{5a}{2h}$$

The mass passing through unit surface in unit time is

$$\rho\bar{\xi} = \rho\left(\frac{c}{2h} + \frac{5g}{4h^2}\right)$$

If the heat conduction is not associated with any mass motion, then we must have

$$c = -\frac{5g}{2h}$$

If we denote the absolute temperature by T, with B a constant, then

$$T = \frac{M}{2}(\overline{\xi^2} + \overline{\eta^2} + \overline{\zeta^2}) \cdot B = \frac{3MB}{h}\left(1 + \frac{ax}{h}\right)$$

hence, neglecting infinitesimal terms, we have

$$\frac{dT}{dx} = \frac{a}{h}T$$

The amount of kinetic energy passing through unit surface in unit time is

$$L = \frac{\rho}{2}(\overline{\xi^3} + \overline{\xi\eta^2} + \overline{\xi\zeta^2})$$

The mean value can easily be calculated by using the assumed value of f. One obtains, after computing all the integrals that arise of the form

$$\iiint \xi^2(\xi^2 + \eta^2 + \zeta^2)^n e^{-h(\xi^2 + \eta^2 + \zeta^2)}\,d\xi d\eta d\zeta$$

which can best be done by differentiating $N\overline{\xi^2}$ with respect to h

$$L = \frac{5}{8}\frac{\rho g}{h^3} = \frac{5MNg}{8h^3}$$

The heat conduction constant is

$$C = -\frac{L}{dT/dx} = -\frac{5}{8}\frac{MNg}{h^3} \cdot \frac{h}{aT} = \frac{5}{16h^2 T A_2 k_1}$$

Noting that

$$\frac{p}{\rho} = \overline{\xi^2} = \frac{1}{2h}$$

one obtains finally

$$C = \frac{5p^2}{4\rho^2 T A_2 k_1}$$

Since I consider the gas molecules to be simple mass-points, Maxwell's parameter β is equal to 1 and the ratio of specific heats is $\gamma = 1\frac{2}{3}$. If we denote the specific heat (of unit mass of the gas) at constant volume in the usual thermal units by w, and the mechanical equivalent of heat by $1/J$, then according to a well-known formula

$$(\gamma - 1)w = \tfrac{2}{3}w = \frac{pJ}{\rho T}$$

The heat conduction constant, measured in the usual thermal units, is therefore

$$C' = JC = \frac{5wp}{6kA_2\rho} = \tfrac{5}{2}w\mu$$

where μ is the viscosity constant. This value of the heat conduction constant is $\frac{3}{2}$ times as large as the one found by Maxwell, because of Maxwell's error in deriving his equation (43) from (39).

The gas molecules have been assumed to be simple mass-points, since with this assumption all the calculations can be carried out exactly. This assumption is clearly not fulfilled in nature, so that the above formulae require some modification if they are to be compared with experiments. If one includes the intramolecular motion following Maxwell's method, he obtains

$$C = \frac{5\beta p^2}{4\rho^2 T A_2 k}, \quad C = \tfrac{5}{2}\,w\mu$$

However, this seems very arbitrary to me, and if one includes intramolecular motion in some other way, he can easily obtain significantly different values for the heat conduction constant. It appears that an exact calculation of this constant from the theory is impossible until we know more about the intramolecular motion. Since the heat conduction constant, whose value was previously thought to be incapable of experimental measurement, has been determined so exactly by Stefan,† it appears that our

---

† [J. Stefan, *Wien. Ber.* **65**, 45 (1872); see also *Wien. Ber.* **47**, 81, 327 (1863); **72**, 69 (1876).]

experimental knowledge is here much better than our theoretical knowledge.

It seems unnecessary to explain what would happen when the function f is not a linear function of x, or when at the same time there takes place heat conduction or motion in other directions.

IV. Treatment of polyatomic gas molecules

We have assumed up to now that each molecule is a single mass-point. This is certainly not the case for the gases existing in nature. We would clearly come closer to the truth if we assumed that each molecule consists of several mass-points (atoms). The properties of such polyatomic gas molecules will be considered in the present section. Note that the previous definitions of symbols are not necessarily retained in this section.

Let the number of mass-points or atoms in a molecule be r. These may be held together by any type of force; we assume only that the force between two points depends only on the distance between them, and acts along the line between them, and that the force is such that the atoms of a given molecule can never be completely separated from each other. I will call this force the internal force of the molecule. During by far the largest part of the time, these forces act only among the atoms in the same molecule. Only when the molecule comes very close to another one will its atoms act on those of the other molecule, and conversely. I call this process, during which two molecules are so close that their atoms act on each other significantly, a collision; and the force which the atoms of different molecules exert on each other will be called the collision force. I assume that this force is also a function of distance that acts along the line of centres, and that the atoms of the two molecules are not exchanged in a collision, but rather that each molecule consists of the same atoms after the collision as before. In order to be able to define precisely the instant when the collision begins, I assume the interaction of two molecules begins whenever the distance between their centres of gravity is equal to a certain quantity l. This distance will then become smaller than l, then increase and when it is again equal to l the collision ends. Of course in reality the moment of beginning a collision is probably not so sharply defined. Our conclusions are not changed if collisions are defined as in my previous paper.† I will therefore retain the above assumption, which is otherwise not inferior in generality to any other assumption, since we have merely assumed that as long as the distance of centres is greater than l, no interaction takes place. If the distance is equal to l, then in many cases the interaction will still be zero, and will not start until later.

In order to define the state of a molecule at a certain time t, we consider three fixed perpendicular directions in space. Through the point at which the centre of our molecule finds itself at time t, we draw three rectangular coordinate axes parallel to these three directions, and we denote the coordinates of the molecule with respect to these axes by $\xi_1, \eta_1, \zeta_1, \xi_2, \ldots, \zeta_r$. Let c_1 be the velocity of the first atom, and u_1, v_1, w_1 its components in the directions of the coordinate axes; the same quantities for the second atom will be c_2, u_2, v_2, w_2; for the third, c_3, u_3, v_3, w_3, and so forth. Then the state of our molecule at time t will be completely determined when we know the values of $6r-3$ quantities

$$\xi_1, \eta_1, \zeta_1, \xi_2, \ldots \xi_{r-1}, \eta_{r-1}, \zeta_{r-1}, u_1, v_1, w_1, u_2, \ldots w_r \quad \text{(A)}$$

ξ_r, η_r, ζ_r are functions of the other ξ, η, ζ, since the centre of gravity is the origin of coordinates. The coordinates of the centre of gravity of our molecule with respect to the fixed coordinate axes do not determine its state but only its position. When our molecule is not interacting with any others, then only the internal

---

† L. Boltzmann, *Wien. Ber.* **63**, 397 (1871).

force between the atoms is present. We can therefore establish, between the time and the $6r-3$ quantities (A), as many differential equations, which we shall call the equations of motion of the molecule. These equations will have $6r-3$ integrals, through which the values of the variables (A) can be expressed as functions of time and of the values of these quantities at some initial time. If we eliminate the time, there remain $6r-4$ equations with the same number of arbitrary constants of integration. Let these be

$$\phi_1 = a_1, \ \phi_2 = a_2, \ldots, \ \phi_\rho = a_\rho$$

where the a's are constants of integration while the ϕ's are functions of the variables (A). ρ is equal to $6r-4$. We can therefore express all but one of the variables (A) as functions of this one variable and the $6r-4$ constants of integration. I will always call this one variable x; it can be either one of the ξ, η, ζ, or one of the u, v, w. As long as the molecule does not collide with another one, these variables (A) satisfy the equations of motion of the molecule, and hence the a's remain constant and the value of each of the variables (A) depends only on the value of x. I will say that a_1, a_2, . . ., a_ρ determine the kind of motion of the molecule, while x determines the phase of the motion. As long as the molecule does not collide with another one, only the variable x determining the phase will change. But when the molecule does collide with another one, then the a's change their values; the kind of motion of the molecule also changes.

We shall now assume that we again have a space $\mathscr{R}$ containing a large number of molecules. All these molecules are equivalent— i.e. they all consist of the same number of mass-points, and the forces acting between them are identical functions of their relative distances for all the molecules. If we now choose, some- where in the space $\mathscr{R}$, a smaller space of volume R, still large compared to the distance between two molecules, then there will be RN molecules in this smaller space. Of these,

$$Rf(t, a_1, a_2 \ldots a_\rho)da_1 da_2 \ldots da_\rho$$

will be in such a state at time t that

$$\left.\begin{array}{l} \phi_1 \text{ lies between } a_1 \text{ and } a_1 + da_1 \\[2mm] \phi_2 \text{ lies between } a_2 \text{ and } a_2 + da_2 \ldots \end{array}\right\} \quad \text{(B)}$$

The constants a determine the kind of motion of a molecule; hence, if the function f is given, then the number of molecules in each of the various kinds of motion at time t in the space R is determined. Hence, we say that the function f determines the distribution of various kinds of motion among the molecules at time t. I assume again that there is already at the initial time, and hence at all subsequent times, a uniform distribution; i.e. the function f is independent of the position of the space R as long as that space is very large compared to the average distance between two neighbouring molecules. For the sake of brevity I will say that a molecule is in a space when its centre of gravity is in that space. We now assume that the value of the function f at time $t = 0$, namely $f(0, a_1, a_2, \ldots)$ is given; we wish to determine the value of f at any later time. The constants a change their values only through collisions; hence f can only change through collisions, and our problem is to establish the equations that determine the variation of f. We must again compute how many collisions occur during a certain time Δt such that before the collision the a's lie between the limits (B), and also how many occur such that after the collision the a's lie between the limits (B). If we add the first number to $f(t, a_1, a_2 \ldots)$ and subtract the latter from it, then we obtain the number of molecules for which the a's lie between the limits (B) after time Δt, namely the quantity $f(t + \Delta t, a_1, a_2 \ldots)da_1 da_2 \ldots$

We now consider some collision between two molecules; for the first molecule, the a's are assumed to lie between the limits (B) before the collision. For the second,

$$\left.\begin{array}{l} \phi_1 \text{ lies between } a_1' \text{ and } a_1' + da_1' \\[2mm] \phi_2 \text{ lies between } a_2' \text{ and } a_2' + da_2' \text{ and so forth} \end{array}\right\} \quad \text{(C)}$$

The collision is not yet completely determined, of course; the phases of the two colliding molecules, as well as their relative position at the beginning of the collision, must also be given. Let the phase of the first molecule be such that

$$x \text{ lies between } x \text{ and } x+dx, \tag{D}$$

while for the second molecule,

$$x \text{ lies between } x' \text{ and } x'+dx' \tag{E}$$

In order to determine the relative position of the two molecules at the beginning of the collision, we denote the angle between the line of centres and the x-axis by θ, the angle between the xy-plane and a plane through the x-axis parallel to the line of centres by ω, and assume that at the beginning the collision

$$\left. \begin{array}{l} \theta \text{ lies between } \theta \text{ and } \theta+d\theta \\ \omega \text{ lies between } \omega \text{ and } \omega+d\omega \end{array} \right\} \tag{F}$$

All collisions that take place in such a way that the conditions (B), (C), (D), (E) and (F) are satisfied, I will call collisions of type (G). The next question is, how many collisions of type (G) occur in unit volume during a certain time Δt? We make the assumption that the internal motions are so rapid, and the collisions so infrequent, that a molecule passes through all its possible phases of motion between one collision and the next. We can then choose Δt so large that each molecule goes through all possible phases of motion during Δt, yet so small that only a few collisions take place during Δt, so that f changes only slightly. We consider some particular molecule whose kind of motion lies between the limits (B); we call it the B molecule. We assume that it passes through all possible phases several times during the time Δt; it can then be shown that the sum of all times during which it has the phase (D) in the course of time Δt is to the total time Δt in the same ratio as $s\,dx$ to $\int s\,dx$, so that the sum of all these times is

$$\tau = \Delta t \frac{s\,dx}{\int s\,dx} \tag{48}$$

where s is given by the following equation:

$$\frac{1}{s} = \sum \pm \frac{\partial a_1}{\partial \xi_1} \frac{\partial a_2}{\partial \eta_1} \cdots \frac{\partial a_\rho}{\partial w_r}$$

The integration is over all possible values of x, i.e. all possible phases. The product $s\ dx$ is always to be taken with the positive sign.

In the functional determinant there will occur the derivatives with respect to all the variable ξ_1, η_1, $\ldots \zeta_{r-1}$, u_1, v_1, $\ldots w_r$ except for x; it can therefore be expressed as a function of x and of the integration constants a. The theorem just stated can be

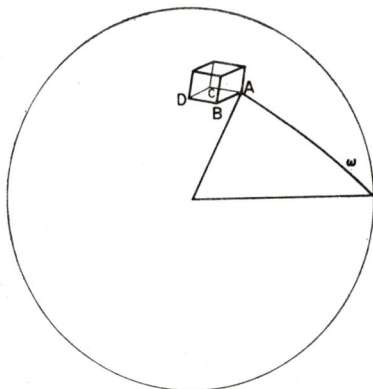

Fig. 4.

shown to be completely similar to Jacobi's principle of the last multiplier;[†] since I have already given the proof in a previous paper,[‡] I will not repeat it here. I have also shown that the situation is not changed when one variable is no longer sufficient to determine the phase.

The total length of all paths which the centre of gravity of (B) traverses relative to the centre of a molecule with property (C),

---

† L. Boltzmann, *Wien. Ber.* **63**, 679 (1871).

‡ See especially note 1 of L. Boltzmann, *Wien. Ber.* **63**, 679 (1871).

during the time interval denoted above by τ, is equal to $\gamma\tau$, where γ is the relative velocity of the centres of the two molecules. If we now describe around the centre of the B molecule a sphere of radius l, then the set of all points of this sphere for which the angles θ and ω lie between the limits (F) form an infinitesimal rectangle $ABCD$ (see Figure 4) of surface area $l^2 \sin \theta \, d\theta \, d\omega$. If we imagine that this rectangle is rigidly connected to the centre of the B molecule, then the total length of the paths which it traverses during τ relative to the C molecule is likewise $\gamma\tau$. All these paths make the same angle with the coordinate axes (neglecting infinitesimal quantities); hence, they also make the same angle (ε) with the plane of the rectangle $ABCD$, since the velocities of all the atoms in B are enclosed within infinitesimally near limits by the conditions (B) and (D). The total volume swept out by our rectangle during τ, if it moves with its relative velocity with respect to C, is

$$V = l^2 \sin \theta d\theta d\omega \,.\, \sin \varepsilon \,.\, \gamma\tau \tag{49}$$

and it is easily seen that all molecules inside this volume will collide with the B molecule in such a way that conditions (F) are fulfilled. It is now a question of finding how many molecules with the kind of motion (C) and phase (E) will lie in this volume. We know that in unit volume there will be $f(t, a_1', a_2', \ldots)da_1' \, da_2' \ldots$ molecules with the kind of motion (C); hence, since the distribution of kinds of motion is uniform, there will be $Vf(t, a_1', a_2' \ldots)$ $da_1' \, da_2' \ldots$ in volume V. However, not all of these molecules will have phase (E). Rather, the number that have this phase will be in the same ratio to the total number as the time during which a molecule has phase (E) is to the time during which it passes through all possible phases. As in equation (48) we can write this as the ratio of $s'dx'$ to $\int s'dx'$, where

$$\frac{1}{s'} = \sum \pm \frac{\partial a_1'}{\partial \xi_1'} \frac{\partial a_a'}{\partial \eta_1'} \cdots$$

This number is therefore

$$v = Vf(t, a_1', a_2' \ldots)da_1' da_2' \ldots \frac{s'dx'}{\int s'dx'} \qquad (50)$$

Here we have assumed that, among the molecules with property (C), the different phases during time τ are distributed in the same way as during time Δt, so that there is no particular correlation between the phases (D) and (E) of the two molecules. If the period of vibration of the molecule with property C is not commensurable with that of the molecule with property B, this is obvious. However, if the two periods of vibration for all or a finite number of molecule pairs are commensurable, then this must be assumed as a property of the initial distribution of states, which is then maintained for all subsequent times.

The expression (50) gives us the number of molecules in volume V for which conditions (C) and (E) are satisfied, and we know that all these, and only these will collide during Δt with the B molecules in such a way that conditions (C), (D), (E) and (F) are satisfied. There are $f(t, a_1, a_2 \ldots)da_1 da_2 \ldots$ molecules with property B in unit volume; if we multiply by this number, we obtain $dn = vf(t, a_1, a_2 \ldots)da_1 da_2 \ldots$ for the number of molecule pairs that collide in unit volume during time Δt in such a way that all five conditions (B), (C), (D), (E) and (F) are satisfied; this is therefore the number of collisions of type (G) during this time. Taking account of equations (48), (49) and (50), we obtain for this number the value:

$$\left. \begin{array}{l} dn = \dfrac{f(t, a_1, a_2 \ldots)f(t, a_1', a_2', \ldots)}{\int sdx \int s'dx'} \, l^2 \gamma \sin\theta \sin\varepsilon \, d\theta d\omega \Delta t \\[2ex] \qquad \times sdx da_1 da_2 \ldots s'dx' da_1' da_2' \ldots \end{array} \right\} \quad (51)$$

If the values of the quantities

$$a_1, a_2, \ldots a_1', a_2', \ldots x, x', \theta, \omega \qquad (J)$$

are given at the beginning of the collision, then the nature of the collision is completely characterized; the values of these quantities at the end of the collision can therefore be calculated. These values after the collision will be denoted by the corresponding capital letters. The quantities

F

$$A_1, A_2, \ldots A'_1, A'_2, \ldots X, X', \Theta, \Omega \qquad \text{(K)}$$

can therefore be expressed as functions of the variables (J). I will denote the set of conditions (B), (C), (D), (E) and (F) as the conditions (G). All collisions for which the values of the variables at the beginning of the collision satisfy (G) will proceed in a very similar way. Hence, at the end of the collisions the values of the variables will also lie between certain infinitesimally near limits. We will assume that for all these, and only these, collisions, the variables after the collision lie between the limits

$$\left.\begin{array}{l} A_1 \text{ and } A_1+dA_1, \; A_2 \text{ and } A_2+dA_2, \ldots A'_1 \text{ and } A'_1+dA'_1 \\ A'_2 \text{ and } A'_2+dA'_2, \ldots X \text{ and } X+dX, \; X' \text{ and } X+dX' \\ \qquad \Theta \text{ and } \Theta + d\Theta, \; \Omega \text{ and } \Omega + d\Omega \end{array}\right\} \quad \text{(H)}$$

Since the variables (K) are functions of the variables (J), we can replace differentials of (J) in equation (51) by differentials of (K); for example, in place of the four differentials dx, dx', $d\theta$, and $d\omega$ we can use differentials of A_1, A_2, A'_1 and A'_2; note that we do not necessarily have to use the four quantities A_1, A_2, A'_1 and A'_2, but rather we could just as well have chosen instead four other variables from (K). We then obtain

$$\left.\begin{array}{l} dn = \dfrac{f(a_1, a_2 \ldots)f(a'_1, a'_2 \ldots)}{\int s\, dx . \int s'\, dx'} ss's_1 l^2\gamma \sin\theta \sin\varepsilon . \Delta t \\ \qquad\qquad \times da_1 da_2 \ldots da_\rho da'_1 \ldots da'_\rho dA_1 dA_2 dA_3 dA_4 \end{array}\right\} \quad (52)$$

where

$$\frac{1}{s_1} = \sum \pm \frac{\partial A_1}{\partial x} . \frac{\partial A_2}{\partial x'} . \frac{\partial A'_1}{\partial \theta} . \frac{\partial A'_2}{\partial \omega}$$

This is the number of collisions that take place in unit volume during time Δt such that before the collision the quantities $\phi_1, \phi_2 \ldots \phi'_\rho$ lie between the limits (B) and (C), and after the collision A_1, A_2, A'_1 and A'_2 lie between

$$A_1 \text{ and } A_1+dA_1 \ldots A'_2 \text{ and } A'_2+dA'_2$$

If we keep $a_1, a_2 \ldots a_\rho$ fixed in this expression and integrate over $a'_1, a'_2 \ldots a'_\rho, A_1, A_2, A'_1$ and A'_2, then we obtain all the collisions that occur such that the a's for one of the colliding molecules lie between the limits (B) before the collision, with no other restriction. This number is therefore:

$$dn' = da_1 da_2 \ldots da_p \Delta t \iint \ldots \left. \begin{array}{c} \dfrac{f(t, a_1, a_2 \ldots) f(t, a_1', a_2' \ldots)}{\int s dx \cdot \int s' dx} \\ \times ss' s_1 l^2 \gamma \sin \theta \sin \varepsilon \, da_1' da_2' \ldots dA_1' dA_2' \end{array} \right\} \quad (53)$$

Each such collision decreases by one the number of molecules for which the a's lie between the limits (B), so we have to subtract the number dn' from $f(t, a_1, a_2 \ldots) da_1 da_2 \ldots$ We still have to add the number of collisions in which the a's lie between the limits (B) after instead of before the collision. Let this number be dN'. Then

$$f(t, a_1, a_2 \ldots) da_1 da_2 \ldots da_p - dn' + dN'$$

is the number of molecules for which the a's lie between the limits (B) at time $t + \Delta t$. Therefore

$$\left. \begin{array}{c} f(t, a_1, a_2 \ldots) da_1 da_2 \ldots da_p - dn' + dN' \\ = f(t + \Delta t, a_1, a_2 \ldots) da_1 da_2 \ldots da_p \end{array} \right\} \quad (54)$$

The number dN' can be found in the following way. We obtained the expression (51) for the number of collisions such that the conditions (G) are satisfied before the collision. For all these collisions, the variables satisfy the conditions (H) at the end of the collision. We merely have to interchange the small and large letters in the expression in order to arrive at the number of collisions for which the values of the variables satisfy at the beginning of the collision conditions that are completely identical with (H) except that the positions of the centres of the two molecules are interchanged, and their line of centres has the opposite direction. The latter number is therefore

$$dN = \left. \begin{array}{c} \dfrac{f(t, A_1, A_2 \ldots) f(t, A_1', A_2' \ldots)}{\int S dX \int S' dX'} SS' l^2 \Gamma \sin \Theta \sin E \\ \times d\Theta d\Omega \Delta t dX dA_1 dA_2 \ldots dA_p dX' dA_1' dA_2' \ldots dA_p' \end{array} \right\} \quad (55)$$

where $S, S', E \ldots$ are constructed from $s, s', \varepsilon \ldots$ by interchanging the variables (J) with the variables (K). For all these collisions, however, the values of the variables at the end of the collision will lie between the limits (G). For it is clear that a collision at the beginning of which the conditions (G) are satisfied will proceed in just the opposite way to one at the beginning of which conditions similar to (H) are satisfied. Whereas for the former collisions the conditions (H) are satisfied at the end, for the latter collisions, conversely, conditions similar to (G) are satisfied at the end. Equation (55) therefore gives us the number of collisions at the end of which the values of the variables satisfy the conditions (G).

We next replace the differentials of X, X', Θ and Ω by differentials of a_1, a_2, a_1' and a_2', so that we obtain

$$dN = \frac{f(t, A_1, A_2 \ldots)f(t, A_1', A_2' \ldots)}{\int SdX \int S'dX'} SS'S_1 l^2 \Gamma \sin \Theta \sin E$$
$$\times \Delta t dA_1 dA_2 \ldots dA_\rho dA_1' \ldots dA_\rho' da_1 da_2 da_1' da_2'$$

In this formula we have

$$\frac{1}{S_1} = \pm \frac{\partial a_1}{\partial X} \cdot \frac{\partial a_2}{\partial X'} \cdot \frac{\partial a_1'}{\partial \Theta} \cdot \frac{\partial a_2'}{\partial \Omega}$$

where, in forming the partial derivatives in the functional determinant, the a's must be considered to be expressed as functions of the independent quantities (G). If we now introduce, in place of the differentials of $A_3, A_4 \ldots A_\rho, A_3' \ldots A_\rho'$ those of $a_3, a_4 \ldots a_\rho, a_3' \ldots a_\rho'$, then we obtain:

$$\left. \begin{array}{l} dN = \frac{f(t, A_1, A_2 \ldots)f(t, A_1', A_2' \ldots)}{\int SdX \int S'dX'} SS'S_1 \sigma l^2 \Gamma \sin \Theta \sin E \\ \qquad \times \Delta t da_1 da_2 \ldots da_\rho' dA_1 dA_2 dA_1' dA_2' \end{array} \right\} \quad (56)$$

where

$$\sigma = \sum \pm \frac{\partial A_3}{\partial a_3} \cdot \frac{\partial A_4}{\partial a_4} \ldots \frac{\partial A_\rho'}{\partial a_\rho'}$$

In this functional determinant, $A_3, A_4 \ldots A_\rho'$ are to be considered as functions of the independent variables $a_1, a_2 \ldots a_\rho', A_1, A_2, A_1', A_2'$. In the expression (56) we shall assume that all quantities are expressed as functions of

$$a_1, a_2 \ldots a_\rho, a_1', \ldots a_\rho', A_1, A_2, A_1', A_2'$$

Furthermore, we consider $a_1, a_2 \ldots a_\rho$ as fixed and integrate over $a_1', a_2' \ldots A_2'$. We thereby obtain the number of collisions after which one of the colliding molecules has values of the a's between the limits (B), while everything else is arbitrary; thus we obtain precisely the quantity called dN'. Therefore

$$dN' = da_1 da_2 \ldots da_\rho \Delta t \iint \ldots \frac{(t, A_1 \ldots)f(t, A_1' \ldots)}{\int S dX \int S' dX'} \left.\right\}$$
$$SS'S_1\sigma l^2 \times \Gamma \sin\Theta \sin E\, da_1' da_2' \ldots da_\rho' dA_1 dA_2 dA_1' dA_2' \tag{57}$$

We shall now expand the right-hand side of equation (54) in powers of Δt and retain only terms of first order. It then reduces to

$$\frac{\partial f(t, a_1, a_2 \ldots)}{\partial t} da_1 da_2 \ldots da_\rho \Delta t = -dn' + dN' \tag{58}$$

Before I substitute here the values of dn' and dN', I will recall a relation which I have already proved in a previous paper.† In our present notation that equation reads:

$$\gamma \sin\theta \sin\varepsilon\, d\xi_1 d\eta_1 \ldots d\zeta_{r-1} du_1 dv_1 \ldots dw_r d\theta d\omega$$
$$= \Gamma \sin\Theta \sin E\, d\Xi_1 dH_1 \ldots d\Omega$$

if $\Xi_1, H_1 \ldots$ denote the values of $\xi_1, \eta_1 \ldots$ at the end of the collision. If we next introduce in this equation $a_1, a_2 \ldots a_\rho, x, x'$ in place of $\xi_1, \eta_1, \ldots w_r$, it becomes

$$ss'\gamma \sin\theta \sin\varepsilon\, da_1 da_2 \ldots da_\rho' dx dx' d\theta d\omega$$
$$= SS'\Gamma \sin\Theta \sin E\, dA_1 dA_2 \ldots dA_\rho' dX dX' d\Xi d\Omega$$

where s and S have the same meanings as before. If we then substitute on the left-hand side dA_1, dA_2, dA_1', dA_2' in place of dx, dx', $d\theta$, $d\omega$, and similarly on the right-hand side, then we obtain

$$ss's_1\gamma \sin\theta \sin\varepsilon\, da_1 da_2 \ldots da_\rho' dA_1 dA_2 dA_1' dA_2'$$
$$= SS'S_1 \Gamma \sin\Theta \sin E\, dA_1 dA_2 \ldots dA_\rho' da_1 da_2 da_1' da_2'$$

If we introduce on the right-hand side of this equation, in place of the differentials of $A_3, A_4 \ldots A_\rho, A_3' \ldots A_\rho'$ those of

† L. Boltzmann, *Wien. Ber.* **63**, 397 (1871).

$a_3, a_4 \ldots a_\rho, a'_3 \ldots a'_\rho$, then we can divide through by all the differentials and we have left:

$$ss's_1\gamma \sin\theta \sin\varepsilon = SS'S_1\Gamma\sigma \sin\Theta \sin E \ldots \qquad (59)$$

We wish to substitute the values (53) and (57) for dn' and dN' into equation (58). If we divide through by $da_1 da_2 \ldots da_\rho$, combine the two integrals on the right into a single one, and finally take account of equation (59), then we obtain

$$\frac{\partial f(t, a_1, a_2 \ldots a_\rho)}{\partial t}$$

$$\left. = \int\int \cdots \left[\frac{f(t, A_1, A_2 \ldots)f(t, A'_1, A'_2 \ldots)}{\int SdX \int S'dX'} \right. \right.$$

$$\left. \left. - \frac{f(t, a_1, a_2 \ldots)f(t, a'_1, a'_2 \ldots)}{\int sdx \int s'dx'} \right] \right\} \qquad (60$$

$$\times ss's_1 l^2\gamma \sin\theta \sin\varepsilon \, da'_1 da'_2 \ldots da'_\rho dA_1 dA_2 dA'_1 dA'_2$$

and this is to be regarded as the basic equation which determines the variation of the function $f(t, a_1, a_2 \ldots)$ with increasing time. We are to assume that all quantities are expressed as functions of the $2\rho+4$ quantities $a_1 a_2 \ldots a_\rho \ldots a'_1 \ldots a'_\rho \; A_1 A_2 A'_1 A'_2$, which completely determine the collision (aside from its absolute position in space). $\int SdX$ is the quantity which one obtains when he expresses $\int sdx$ as a function of these $2\rho+4$ variables; then by permuting the small and large letters, and again expressing the $A_3 A_4 \ldots$ as functions of the $2\rho+4$ variables, one similarly obtains $\int SdX'$.

One sees at once that this variation is equal to zero whenever the expression in brackets vanishes for all values of the variables; for then the integral must also vanish, and this is nothing but the result that I have already found in a different form in the paper on thermal equilibrium already cited. It remains only to prove that the integral can only vanish when the integrand is everywhere zero. Otherwise it might happen that the integral vanishes because the integrand is sometimes positive and sometimes negative. In

order to obtain this latter proof, we proceed in just the same way
as before, in the theory of the properties of monatomic molecules.
Since the functional determinant s is only a function of x, a_1,
$a_2 \ldots a_\rho$, and the limits of the integral $\int sdx$ depend only on the
a's, it follows that the latter integral is a function of $a_1, a_2 \ldots a_\rho$.
Hence, $f(t, a_1, a_2 \ldots)/ \int sdx$ is a function of $t, a_1, a_2 \ldots a_\rho$, and we
shall set:

$$\frac{f(t, a_1, a_2 \ldots a_\rho)}{\int sdx} = \phi(t, a_1, a_2 \ldots a_\rho) \tag{61}$$

Furthermore, for brevity we set

$$ss's_1 l^2 \gamma \sin\theta \sin\varepsilon = p$$

Then we can write equation (13) in the somewhat shorter form:

$$\frac{\partial \phi(t, a_1, a_2 \ldots)}{\partial t} \int sdx = \iint \ldots [\phi(t, A_1, A_2 \ldots)\phi(t, A_1', A_2' \ldots)$$
$$- \phi(t, a_1, a_2 \ldots)\phi(t, a_1', a_2' \ldots)]pda_1'da_2' \ldots da'dA_1 dA_2 dA_1' dA_2' \tag{62}$$

The quantity p has two properties that will be used later. First,
it is symmetric with respect to the variables $a_1, a_2 \ldots a_\rho A_1 A_2$
and $a_1' a_2' \ldots a_\rho' A_1' A_2'$, i.e. it does not change its value when one
simultaneously interchanges a_1' and a_1, a_2' and a_2, and so forth.
For this means simply that one ascribes to the second of the
colliding molecules the state that the first had before the permutation, and conversely; and it is clear that this does not change the
relative velocities, etc., of the two molecules. That the product ss'
and the quantity s_1 are likewise symmetric in this way can be
verified by simply looking at the expressions for these quantities.

Second, p is necessarily positive; for the number dn represented
by equation (52) must be positive, and this expression for dn
includes, aside from p, only positive factors (we assume that the
differentials are always positive in any case); consequently p must
likewise be positive. If we now set

$$E = \iint \ldots \phi(t, a_1, a_2 \ldots).\int sdx.\log\phi(t, a_1, a_2 \ldots)da_1 da_2 \ldots da_\rho \tag{63}$$

where " log " means the natural logarithm, and the integration is to be extended over all possible values of the a's, so that E is a function only of the time. Since the limits of the integrals that determine E do not depend on t, we can find dE/dt by differentiating with respect to t under the integral sign, and then only insofar as the quantities under the integral sign contain t explicitly; for the a's are just variables over which we integrate. We know that $\int sdx$ does not depend on time; only $\phi(t, a_1, a_2 \ldots)$ contains t. If we take account of the fact that

$$\iint \ldots \phi(t, a_1, a_2 \ldots) . \int sdx . da_1 da_2 \ldots da_\rho$$

is the total number of molecules in unit volume, so that its time-derivative is equal to zero, then we obtain:

$$\frac{dE}{dT} = \iint \ldots \frac{\partial \phi(t, a_1, a_2 \ldots)}{\partial t} \log \phi(t, a_1, a_2 \ldots) . \int sdx . da_1 da_2 \ldots da_\rho$$

If we substitute here for $\partial \phi(t, a_1, a_2 \ldots)/\partial t$ its value from equation (62), and write, for brevity,

$$\phi(t, a_1, a_2 \ldots) = \phi; \quad \phi(t, a_1', a_2' \ldots) = \phi'; \quad \phi(t, A_1, A_2 \ldots) = \Phi;$$
$$\phi(t, A_1', A_2' \ldots) = \Phi'$$

then we find

$$\frac{dE}{dt} = \iiint \ldots \log \phi . [\Phi\Phi' - \phi\phi'] \, p \, da_1 da_2 \ldots da_\rho da_1' \left.\begin{array}{l} \\ \\ \ldots da_\rho' dA_1 dA_2 dA_1' dA_2' \end{array}\right\} \quad (64)$$

In the definite integral on the right-hand side of (64) we may again label the variables over which we integrate just as we please. We can therefore, for example, interchange $a_1 a_2 \ldots A_2$ and $a_1' a_2' \ldots A_2'$. This does not change either p or the quantity in brackets, but only changes $\log \phi$ to $\log \phi'$. We therefore find, if we also reverse the order of integration, that the variables are integrated in the same order as before:

$$\frac{dE}{dt} = \iint \ldots \log \phi' [\Phi\Phi' - \phi\phi'] \, p \, da_1 da_2 \ldots da_\rho' dA_1 dA_2 dA_1' dA_2'$$

$$(65)$$

We shall now interchange those a's which are denoted in (64) by small Latin letters with the capitals, so that we obtain:

$$\frac{dE}{dt} = \iint \dots \log \Phi[\phi\phi' - \Phi\Phi'] P dA_1 dA_2 \dots dA'_\rho da_1 da_2 da'_1 da_2$$

(66)

where

$$P = SS'S_1 l^2 \Gamma \sin \Theta \sin E$$

One therefore obtains P by expressing p as a function of the $2\rho + 4$ quantities $a_1 a_2 \dots a'_\rho \; A_1 \dots A'_2$ and then interchanging small and large letters. However, we know that $A_3 A_4 \dots A_\rho$ $A'_3 \dots A'_\rho$ can be expressed as functions of $a_1 a_2 \dots a'_\rho A_1 A_2 A'_1$ and A'_2. We can therefore replace the differentials of $A_3 A_4 \dots A_\rho$ $A'_3 A'_4 \dots A'_\rho$ in equation (66) by those of $a_3 a_4 \dots a_\rho a'_3 \dots a'_\rho$, which again is only a purely formal change. Indeed, we may replace under the integral sign those variables over which we integrate by any arbitrary functions of the same variables, and thus also, if we wish, by those which we earlier called $a_3 a_4 \dots A_3 A_4$. Equation (66) is thereby transformed to

$$\frac{dE}{dt} = \iint \dots \log \Phi[\phi\phi' - \Phi\Phi'] P\sigma da_1 da_2 \dots da'_\rho dA_1 dA_2 dA'_1 dA'_2$$

or, if we take account of equation (59), to

$$\frac{dE}{dt} = \iint \dots \log \Phi[\phi\phi' - \Phi\Phi'] p da_1 da_2 \dots da'_\rho dA_1 dA_2 dA'_1 dA'_2$$

(67)

In all these equations the integrations are over all possible values of the variables.

If we finally interchange $a_1 a_2 \dots A_2$ and $a'_1 a'_2 \dots A'_2$, we find:

$$\frac{dE}{dt} = \iint \dots \log \Phi'[\phi\phi' - \Phi\Phi'] p da_1 da_2 \dots da'_\rho dA_1 dA_2 dA'_1 dA'_2$$

(68)

F*

If we now add together equations (64), (65), (67) and (68) and divide by 4, we arrive at the result

$$\frac{dE}{dt} = \frac{1}{4} \int \int \dots \log \left(\frac{\phi\phi'}{\Phi\Phi'} \right) \cdot [\Phi\Phi' - \phi\phi'] p \, da_1 da_2 \dots dA_2'$$

Since p is always positive, it follows from this expression immediately that dE/dt can never be positive, and therefore E itself can only decrease or remain constant. The latter can only be the case when the expression in brackets,

$$\left. \begin{aligned} &\phi(t, a_1, a_2 \dots a_\rho) \cdot \phi(t, a_1', a_2' \dots a_\rho') \\ &\quad - \phi(t, A_1, A_2 \dots A_\rho) \cdot \phi(t, A_1', A_2' \dots A_\rho') \end{aligned} \right\} \quad (69)$$

vanishes for all values of the variables. Hence, the distribution of states cannot fluctuate periodically between certain limits, and for the limit that it approaches with increasing time, the expression (69) must vanish in general.

The meaning of the above transformation of equation (64) can be further clarified, as in the case of monatomic gas molecules, by breaking up the integral into a sum. We set

$$a_1 = b_1 \varepsilon, \quad a_2 = b_2 \varepsilon, \dots, \quad a_\rho' = b_\rho' \varepsilon, \quad A_1 = B_1 \varepsilon \dots$$

where ε is some very small quantity, and the b's are integers. Further:

$$\phi(t, a_1, a_2 \dots) = w_{b_1 b_2 \dots}, \quad \int s \, dx = v_{b_1 b_2 \dots},$$
$$\varepsilon^{\rho+4} p = D_{B_1 B_2 \dots B_1' B_2'}^{b_1 b_2 \dots b_1' b_2'} \dots$$

Since D depends on all the variables, it must be given $2\rho + 4$ indices: the rest of the indices must be attached to it by symmetry. The v's are constants, whose values in general depend on the indices; the w's are functions of time. The system of ordinary differential equations, which now replaces equation (64), is the following:

$$\left. \begin{aligned} v_{b_1 b_2 \dots} \frac{dw_{b_1 b_2 \dots}}{dt} = \Sigma D_{B_1 B_2 \dots B_1' B_2'}^{b_1 b_2 \dots b_1' b_2'} \big[w_{B_1 B_2} \dots w_{B_1' B_2'} \dots \\ - w_{b_1 b_2} \dots w_{b_1' b_2'}^{b_1' b_2'} \dots \big] \end{aligned} \right\} \quad (70)$$

The summation is to be taken over all possible values of $b_1'b_2' \ldots B_1B_2 \ldots$ In equation (70) itself, $b_1b_2 \ldots b_\rho$ can have many different values; it represents a system of many differential equations. Equation (59) now reads, in our present notation:

$$D_{B_1B_2 \ldots B_1'B_2' \ldots}^{b_1b_2 \ldots b_1'b_2' \ldots} = D_{B_1'B_2' \ldots B_1B_2}^{b_1'b_2' \ldots b_1b_2} = D_{b_1b_2 \ldots b_1'b_2' \ldots}^{B_1B_2 \ldots B_1'B_2' \ldots}$$

and one easily finds from this equation and from the system (70) that the derivative of

$$\Sigma v_{b_1b_2} \ldots w_{b_1b_2} \ldots \log w_{b_1b_2} \ldots$$

can never be positive. The summation in this last expression is over all possible values of $b_1b_2 \ldots b_\rho'$. This quantity therefore decreases continually until

$$w_{b_1b_2} \ldots w_{b_1'b_2'} \ldots = w_{B_1B_2} \ldots w_{B_1'B_2'} \ldots$$

for all possible values of $b_1b_2 \ldots b_\rho'$ and for all values of $B_1B_2B_1'B_2'$ consistent with these. For D cannot be zero for any group of indices that correspond to a possible state of the molecule. If it were, then some collision would have probability zero.

We can now rewrite the quantity E given by equation (63). We saw that

$$f(t, a_1, a_2 \ldots)da_1da_2 \ldots da_\rho \left(\frac{sdx}{\int sdx}\right)$$
$$= \phi(t, a_1, a_2 \ldots a_\rho)da_1da_2 \ldots da_\rho.sdx$$

is the number of molecules in unit volume at time t for which conditions (B) and (D) are satisfied. If we replace the differentials of $a_1a_2 \ldots a_\rho$ by those of $\xi_1 \eta_1 \ldots \zeta_{r-1} u_1v_1 \ldots w_r$, then this expression becomes

$$\phi(t, a_1, a_2 \ldots a_\rho)d\xi_1d\eta_1 \ldots d\zeta_{r-1}du_1 \ldots dw_r$$

If we again suppose that $a_1a_2 \ldots a_\rho$ are expressed in terms of $\xi_1\eta_1 \ldots w_r$, then ϕ will be a function of the latter variables. $\phi(t, a_1, a_2 \ldots a_\rho)$ then reduces to $F(t, \xi_1, \eta_1 \ldots w_r)$, so that therefore

$$F.d\xi_1d\eta_1 \ldots d\zeta_{r-1}du_1 \ldots dw_r$$

is the number of molecules in unit volume for which ξ_1 lies between ξ_1 and $\xi_1 + d\xi_1 \ldots w_r$ between w_r and $w_r + dw_r$. We can then write the expression for E as

$$E = \iint \ldots F \log F \; sdxda_1da_2 \ldots da_\rho$$

or, if we use differentials of $\xi_1, \eta_1 \ldots w_r$,

$$E = \iint \ldots F \log F d\xi_1 d\eta_1 \ldots d\zeta_{r-1} du_1 \ldots dw_1$$

I note that the calculation can be carried out in the same way if several kinds of molecules are present in the same container. If we denote the corresponding quantities for the second kind of molecule by adding a star, and so forth, then the quantity

$$E = \iint \ldots F \log F d\xi_1 d\eta_1 \ldots dw_r$$
$$+ \iint \ldots F^* \log F^* d\xi_1^* d\eta_1^* \ldots dw_r^* \quad (71)$$

can never increase.

V. The molecules do not make a large number of vibrations between one collision and the next

In the previous section I assumed that each molecule makes a very large number of vibrations between one collision and the next. It can easily be shown also that when this is not the case the quantity E defined by equation (71) still cannot increase. It is convenient here to denote the number of molecules in unit volume for which at time t the variables $\xi_1, \eta_1 \ldots w_r$ lie between the limits

$$\xi_1 \text{ and } \xi_1 + d\xi_1 \ldots w_r \text{ and } w_r + dw_r \quad \text{(L)}$$

by $f(t, \xi_1, \eta_1 \ldots w_r) \, d\xi_1 d\eta_1 \ldots dw_r$; f will now be the same function which was called F in the previous section. It can next be proved that the quantity E is not changed by the internal motions of the atoms in the molecules, so that it remains constant if the molecules do not collide with each other. The variables ξ_1, $\eta_1 \ldots w_r$ will be determined by differential equations which we have called the equations of motion of a molecule. Because of these equations of motion, we know that at time $t+\delta t$ the same molecules whose variables were within the limits (L) at time t will now have their variables lying between the limits:

$$\xi_1' \text{ and } \xi_1'+d\xi_1' \ldots w_r' \text{ and } w_r'+dw_r' \qquad \text{(M)}$$

In my paper on thermal equilibrium among polyatomic molecules,† I showed that‡

$$d\xi_1'd\eta_1' \ldots dw_r' = d\xi_1 d\eta_1 \ldots dw_r$$

If no collision occurs, then the same molecules will have their variables within the limits (M) at time $t+\delta t$ as had them within (L) at time t. The numbers of molecules in these two groups will therefore be equal, since both contain exactly the same molecules. But the number in (M) at $t+\delta t$ is

$$f(t+\delta t, \xi_1' \ldots w_r')d\xi_1' \ldots dw_r'$$

the number in (L) at t is

$$f(t, \xi_1 \ldots w_r)d\xi_1 \ldots dw_r$$

Therefore

$$f(t+\delta t, \xi_1' \ldots w_r')d\xi_1' \ldots dw_r' = f(t, \xi_1 \ldots w_r)d\xi_1 \ldots dw_r \quad \text{(72)}$$

(We assume that the distribution of states is uniform everywhere in the gas.) Taking account of equation (72), we obtain

$$f(t+\delta t, \xi_1' \ldots w_r') = f(t, \xi_1 \ldots w_r)$$

therefore we have also

$$f' \log f' \, d\xi_1'd\eta_1' \ldots dw_r' = f \log f \, d\xi_1 \ldots dw_r \qquad \text{(73)}$$

if we write f for $f(t, \xi_1 \ldots w_r)$ and f' for $f(t+\delta t, \xi_1' \ldots w_r')$. Since equation (73) holds for each set of differentials, we obtain a similar

---

† L. Boltzmann, *Wien. Ber.* **63**, 397 (1871).

‡ If one wishes to calculate with finite quantities, this equation reads:

$$\lim \frac{d\xi_1' \, d\eta_1' \ldots dw_r'}{\delta t \, d\xi_1 d\eta_1 \ldots dw_r} = 1$$

expression if we integrate on both sides over all possible values of the variables. Therefore

$$\iint \ldots f' \log f' \, d\xi_1' \ldots dw_r' = \iint \ldots f \log f \, d\xi_1 \ldots dw_r$$

where we have to integrate over all possible values of the variables on both sides, so that it makes no difference how we label the variables. One can therefore write also:

$$\iint \ldots f(t+\delta t, \xi_1 \ldots w_r) \log f(t+\delta t, \xi_1 \ldots w_r) d\xi_1 \ldots dw_r$$
$$= \iint \ldots f \log f \, d\xi_1 \ldots dw_r$$

The right-hand side of this equation represents the value of E at time $t+\delta t$, while the left is its value at time t. The two are equal. E therefore does not change its value if the atoms in the molecules move according to their equations of motion, as long as the molecules do not collide with each other.

It is now a question of finding how much E changes during collisions. If δt is taken to be very small, then the total variation of E will be the sum of individual variations. If we denote the increase experienced by E as a result of collisions by δE, then we have

$$\delta E = \iint \ldots \log f \, \delta f d\xi_1 \ldots dw_r \qquad (74)$$

where δf is the increase experienced by f during time δt as a result of collisions. Hence, the number of molecules (δN) which attain the state (L) through collisions in time δt is $\delta f d\xi_1 d\eta_1 \ldots dw_r$ greater than the number of molecules (δn) which leaves this state through collisions:

$$\delta N - \delta n = \delta f d\xi_1 d\eta_1 \ldots dw_r \qquad (75)$$

I assume that a collision of two molecules begins when their centres of gravity are at a distance l. The angle between the line of centres and the x-axis will again be called θ; the angle between the xy-plane and a plane parallel to the line of centres passing through the x-axis will be called ω. Then the number of collisions in unit volume during time δt for which at the beginning the variables θ and ω lie between

$$\theta \text{ and } \theta+d\theta, \ \omega \text{ and } \omega+d\omega \qquad (N)$$

and furthermore the variables $\xi_1 \eta_1 \ldots w_r$ for one of the colliding molecules lie between the limits (L), while the variables for the other molecule may lie within any other limits such as

$$\xi_1' \text{ and } \xi_1' + d\xi_1' \ldots w_r' \text{ and } w_r' + dw_r' \tag{P}$$

will be:

$$f d\xi_1 \ldots dw_r f' d\xi_1' \ldots dw_r' g l^2 \sin \theta d\theta d\omega \delta t$$

where g is the relative velocity of the centres of the molecules. For all these, and only these molecules, θ and ω will lie between the limits

$$\Theta \text{ and } \Theta + d\Theta, \ \Omega \text{ and } \Omega + d\Omega \tag{N*}$$

at the end of the collision. Furthermore, the variables ξ_1, $\eta_1 \ldots w_r$ for the first molecule will lie between the limits

$$\Xi_1 \text{ and } \Xi_1 + d\Xi_1 \ldots W_r \text{ and } W_r + dW_r \tag{L*}$$

and for the second molecule, between the limits

$$\Xi_1' \text{ and } \Xi_1' + d\Xi_1' \ldots W_r' \text{ and } W_r' + dW_r' \tag{P*}$$

Each of these collisions will remove a molecule from the state (L). Hence, the total depletion during time δt is

$$\delta n = f d\xi_1 \ldots dw_r \delta t \iint \ldots f' d\xi_1' \ldots dw_r' g l^2 \sin \theta d\theta d\omega \tag{76}$$

The number of collisions for which initially the variables lie between the limits (L*), (N*), and (P*) is

$$\delta v = F d\Xi_1 \ldots dW_r F' d\Xi_1' \ldots dW_r' G l^2 \sin \Theta d\Theta d\Omega \delta t$$

For all these collisions, the variables lie between the limits (L), (N) and (P) afterwards, since these collisions are just the reverse of those considered earlier.† Since the variables

$$\Xi_1 \ldots W_r, \Xi_1' \ldots W_r', \Theta, \Omega \tag{R}$$

are functions of

$$\xi_1 \ldots w_r, \xi_1' \ldots w_r', \ \theta, \omega \tag{Q}$$

---

† Of course in comparing the final state of one collision with the initial state of another one, the positions of the centres of mass of the two molecules have to be interchanged, since at the beginning they move towards each other, and at the end away from each other.

one can introduce the latter variables in place of the former in the expression (76). Then we have

$$G \sin \Theta \, d\Xi_1 \ldots dW_r' d\Theta d\Omega = g \sin \theta \, d\xi_1 \ldots dw_r' d\theta d\omega$$

as follows from the general theorem proved in the second section of a previous paper.† Consequently we have

$$\delta v = FF' d\xi_1 \ldots dw_r' \sin \theta g l^2 d\theta d\omega \delta t \qquad (77)$$

In the above equation we wrote F, F', f', for $f(t, \Xi_1 \ldots W_r)$, $f(t, \Xi_1' \ldots W_r')$ and $f(t, \xi_1' \ldots w_r')$. G is the quantity that one obtains when he interchanges the variables (Q) and (R) in g. In the expression (77) one is supposed to consider that the variables (R) are expressed as functions of the variables (Q). If we integrate $\xi_1' \ldots w_r'$, θ, ω over all possible values, then we obtain the number of molecules which attain the state (L) by collisions in unit time, and thus the number earlier denoted by δN:

$$\delta N = d\xi_1 \ldots dw_r \delta t \iint \ldots FF' g l^2 \sin \theta \, d\xi_1' \ldots dw_r' d\theta d\omega$$

If we substitute these values for δn and δN into equation (75), we obtain

$$\delta f = \delta t \iint \ldots (FF' - ff') g l^2 \sin \theta \, d\xi_1' \ldots dw_r' d\theta d\omega$$

and if we substitute this into equation (74), we find

$$\delta E = \delta t \iint \ldots \log f \, (FF' - ff') g l^2 \sin \theta \, d\xi_1 \ldots dw_r d\xi_1' \ldots dw_r' d\theta d\omega$$

By interchanging the labels of the two colliding molecules, we obtain just as before

$$\delta E = \delta t \iint \ldots \log f' \, (FF' - ff') g l^2 \sin \theta \, d\xi_1 \ldots dw_r d\xi_1' \ldots dw_r' d\theta d\omega$$

By replacing the variables (Q) by (R) and then changing the labels of the variables, we find

$$\delta E = -\delta t \iint \ldots \log F \, (FF' - ff') g l^2 \sin \theta \, d\xi_1 \ldots dw_r' d\theta d\omega$$

$$\delta E = -\delta t \iint \ldots \log F' \, (FF' - ff') g l^2 \sin 0 \, d\xi_1 \ldots dw_r' d\theta d\omega$$

† L. Boltzmann. *Wien. Ber.* **63**, 397 (1871).

(Note that the variables (R) are the same functions of (Q) as the (Q) are of the (R).) Adding together the four equations, we find

$$\delta E = \tfrac{1}{4}\delta t \iint \ldots \log\left(\frac{ff'}{FF'}\right).(FF'-ff')gl^2 \sin\theta\, d\xi_1 \ldots dw_r d\theta d\omega$$

whence it follows again the E can only decrease as a result of collisions, and since it does not change through the motions of atoms in the molecules, it follows that in general it can only decrease.

If the distribution of states at the initial time were not uniform, then f would also depend on the position (x, y, z) in the gas. We would then have instead of E a more general expression. If

$$f(t,x,y,z,\xi_1,\eta_1 \ldots w_r)dxdydzd\xi_1 \ldots dw_r$$

is the number of molecules in the volume element $dxdydz$ at (x, y, z) at time t, for which the variables lie between the limits (L), then the quantity

$$E = \iint \ldots f \log f . dxdydzd\xi_1 \ldots dw_r \qquad (78)$$

cannot increase. In order to prove this, we shall make the problem even more general. Suppose that we have many systems of mass-points (molecules). Each consists of r mass-points $m_1 m_2 \ldots m_r$ (the m's are the actual masses of the points as well as their labels). Let the mass m_1 be the same for all systems; likewise the mass m_2, etc. Let x_1, y_1, z_1 be the coordinates and u_1, v_1, w_1 the velocity components of m_1. Let $x_2, y_2 \ldots$ have similar meanings; and indeed it makes no difference whether the origin of coordinate is the same or different for the different systems. The force that acts on any of the mass-points will be assumed to be a function of the coordinates $x_1, y_1, z_1, x_2 \ldots z_r$ such that a potential function exists, and we assume that this potential is the same function of $x_1 y_1 \ldots z_r$ for all the systems. If we denote by $f(t, x_1, y_1, z_1 \ldots w_r)$ the number of systems for which the variables $x_1, y_1, z_1, x_2 \ldots z_r$ $u_1 \ldots w_r$ lie between the limits

$$x_1 \text{ and } x_1+dx_1 \ldots w_r \text{ and } w_r+dw_r \qquad (S)$$

and set

$$E = \iint \ldots f \log f \, dx_1 dy_1 \ldots dw_r \qquad (79)$$

then it can be proved just as before that E does not change as a result of the motion of the mass-points of the system as long as only the internal forces of that system act between the points. Now assume that there are interactions between the points of different systems, such that the force between two points is a function of their distance and acts in the directions of the line of centres. The conditions shall be such that when the interaction of two systems begins, there is never (or almost never) a third system simultaneously interacting with these two. (For example, the intertion of two systems might occur whenever a point of one of them comes unusually close to a point of the other.) The number of pairs of systems that interact during time δt such that the variables determining the state of one system lie initially within the limits (S), and those of the other between

$$x_1' \text{ and } x_1' + dx_1' \ldots w_r' \text{ and } w_r' + dw_r'$$

is then

$$f(t, x_1 \ldots w_r)f(t, x_1' \ldots w_r')dx_1 \ldots dw_r dx_1' \ldots dw_r' \delta t . \phi \quad (80)$$

ϕ is a function of the relative distances and the velocities of the atoms of the two systems. If one takes account of the fact that this function must have the general property expressed by equation (19) of my paper on the thermal equilibrium of polyatomic molecules,† then one can prove as before that E can only decrease as a result of interactions of different systems. The proof can also be carried through when the systems are not all the same but rather are of two or more kinds, as long as a large number of systems of each kind are present. (One sees at once that a mixture of gas molecules with non-uniform velocity distributions is only a special case of this.) When the potential function has the value gz, one finds for monatomic gas molecules

$$f = Ae^{-h(gz + \frac{1}{2}mc^2)}$$

---

† L. Boltzmann, *Wien. Ber.* **63**, 397 (1871).

The well-known formula for barometric height measurements as well as all aerostatic formulae thus follow as special cases from the formulae for thermal equilibrium.

VI. Solution of equation (81) and calculation of the entropy

We have proved that for polyatomic gas molecules in the case of equilibrium of kinetic energy, the expression (69) must vanish:

$$\phi(a_1, a_2 \ldots)\phi(a_1', a_2' \ldots) - \phi(A_1, A_2 \ldots)\phi(A_1', A_2' \ldots) = 0 \tag{81}$$

must hold for all possible values of the variables

$$a_1 a_2 \ldots a_1' a_2' \ldots A_1 A_2 \ldots A_1' A_2' \ldots A_\rho' \tag{82}$$

We now have to find a function ϕ that satisfies this equation. It is clear that if λ is the total kinetic energy and work contained in a molecule, then the value $\phi = Ae^{-h\lambda}$ (where A and h are constants) must satisfy (81). This is the solution of the problem which I have already found in my previous paper.† It remains only to prove that this is the only possible solution of (81). While this proof was the easiest step in the case of monatomic gas molecules, here it is the most difficult, because it is not actually possible to give in general the various equations which relate the values of the variables before and after the collision. It can still be proved at least in the case of diatomic molecules under the assumption of a particular kind of interaction of the molecules during the collision. We assume that each molecule consists of two atoms. Let r be their distance, and let $m(\chi'(r))/2$ be the force of attraction between them when the molecule is not interacting with another one, so that $m(\chi'(r))/2$ is the potential function. In order not to make the formulae too complicated, I assume that all the atoms have equal mass (the more general case can be treated in a similar way). The sum of the values of the potential and kinetic energies of the two atoms of a molecule remains constant from one collision to the next. This sum divided by half the mass of an atom will be denoted

---

† L. Boltzmann, *Wien. Ber.* **63**, 397 (1871).

by a; four times the square of the velocity of the centre of mass of the molecule will be denoted by b; four times the surface swept out by the radius vector from the centre of mass to one of the atoms in unit time will be denoted by c. Thus a, b, and c are the only constants of integration determining the nature of the atomic path. The others determine only its position in space. One can place a plane through the line of centres of two molecules in such a way that the velocity component w perpendicular to this plane is the same for both atoms. This plane will be called the orbital plane. Let α be the angle it makes with any fixed plane, and let β be the angle between its line of intersection with the fixed plane and a fixed line drawn in the fixed plane. Let γ be the angle which that velocity component of the centre of mass of the molecule which is parallel to the orbital plane makes with the line of intersection of the orbital plane and the fixed plane; let δ be the angle which the line of apsides which the orbital curve of the atom makes with the same line by its motion around the centre of mass. We assume that the gas has the same properties in all directions; between one collision of the molecule and the next, there are many maxima and minima of the distance between its atoms (apside positions). Finally, the angle between two successive apside lines is not a rational fraction of π (with the exception of an infinitesimal number of special path shapes). Then a, b, c, w, α, β, γ, δ are the constants of integration denoted by $a_1 a_2 \ldots a_\rho$ in the previous section, r is equal to 2, $\rho = 6r - 4 = 8$, and the quantity $\phi(a_1, a_2 \ldots a_\rho) da_1 \ldots da$ must have the form

$$\phi(a, b, c, w) \sin \alpha \, da \, db \, dc \, dw \, d\alpha \, d\beta \, d\gamma \, d\delta$$

Because of the equivalence of all directions in space, ϕ cannot depend on the angles α, β, γ. Because of our assumption about the apside lines, all values of δ are equivalent. We shall now consider only those collisions for which w and the orbital planes for the two colliding molecules are identical. Equation (81) must hold for all collisions and hence also for this kind. However, α is not changed by the collision, so that equation (81) reduces to

$$\phi(a, b, c, w)\phi(a', b', c', w') = \phi(A, B, C, w)\phi(A', B', C', w') \quad (83)$$

We shall now make a special assumption about the interaction between two molecules during the collision. The collision of two molecules shall be such that an atom of one molecule rebounds from one of the other molecules like an elastic sphere. (We label the colliding atom as the first one of its molecule.) We now construct, parallel to the line of centres of the colliding atoms at the beginning of the collision, a fixed x-axis (any line of centres shall fall in the orbital plane) and perpendicular to this but parallel to the orbital plane, a fixed y-axis. At the beginning of the collision, let u, v be the velocity components of the first atom of the first of the two colliding molecules in the directions of these two axes; let ξ, η be the coordinates of the same with respect to a system whose origin is at the centre of mass of the molecule, and whose axes are parallel to our fixed ones. Let u_1, v_1 be the velocity components of the second atom of the first molecule. The notation for the other molecule, and for quantities at the end of the collision, will be chosen in the usual way. Then

$$\left.\begin{aligned}
a &= u^2 + v^2 + u_1^2 + v_1^2 + 2w^2 + \chi \\
b &= (u + u_1)^2 + (v + v_1)^2 + 4w^2 \\
a' &= u'^2 + v'^2 + u_1'^2 + v_1'^2 + 2w'^2 + \chi' \\
b' &= (u' + u_1')^2 + (v' + v_1')^2 + 4w'^2
\end{aligned}\right\} \quad (84)$$

The collision reverses the x-components of the velocities of the colliding atoms hence $U = u'$, $U' = u$; all other capital letters have the same values as the corresponding small ones. Therefore

$$\left.\begin{aligned}
A &= u'^2 + v^2 + u_1^2 + v_1^2 + 2w^2 + \chi \\
B &= (u' + u_1)^2 + (v + v_1)^2 + 4w^2 \\
A' &= u^2 + v'^2 + u_1'^2 + v_1'^2 + 2w'^2 + \chi' \\
B' &= (u + u_1')^2 + (v' + v_1')^2 + 4w'^2
\end{aligned}\right\} \quad (85)$$

We shall now show that, if arbitrary values of the quantities

$$a, b, c, a', b', c', w, w' \quad (86)$$

are given, then the quantities $u, v, \xi \ldots$, which are not hereby determined, can also be chosen such that given values of

$$A, B, C, A', B', C'_, \tag{87}$$

will occur after the collision, provided only that the quantities (86) and (87) satisfy the single condition

$$a + a' = A + A' \tag{88}$$

which is just the equation for conservation of energy; thus there are no other relations between the quantities (86 and (87).

We now set, for the sake of brevity,

$$b - a - B + A = g, \quad b' - a' - B' + A' = g'$$

so that g and g' are the given quantities.

We find:

$$g = 2u_1(u - u'), \quad g' = 2u'_1(u' - u)$$

hence

$$u' = u - \frac{g}{2u_1}, \quad u'_1 = -\frac{g'}{g}u_1 \tag{89}$$

and consequently

$$b - B = (u + u_1)^2 - \left(u - \frac{g}{2u_1} + u_1\right)^2 \tag{90}$$

$$b' - B' = \left(u - \frac{g}{2u_1} - \frac{g'}{g}u_1\right)^2 - \left(u - \frac{g'}{g}u_1\right)^2$$

From the two latter equations it follows that

$$b - B + b' - B' = g + g'$$

which is just the equation for conservation of energy. This must be satisfied; one can then chose u_1 at will. The other equations (90) determine u quadratically, while equations (89) determine u' and u'_1. Equations (89) and (90) completely replace four of equations (84) and (85), There remain still the other four. These involve the equations for the c's. They read:

$$c = \xi(v-v_1)-\eta(u-u_1), \quad c' = \xi'(v'-v_1')-\eta'(u'-u_1')$$
$$C = \xi(v-v_1)-\eta(u'-u_1), \quad C' = \xi'(v'-v_1')-\eta'(u-u_1') \qquad (91)$$

whence it follows that

$$\eta = \frac{c-C}{u'-u}, \; \eta' = -\frac{c'-C'}{u'-u}$$

which again replaces two of equations (91), and determines η and η'. There remain still four of equations (84) and (85), and two of equations (91) to be satisfied, thus for example

$$\left.\begin{array}{l} v^2+v_1^2 = a-u_1^2-2w^2-\chi \\[1mm] (v+v_1)^2 = n-(u+u_1)^2-4w^2 \\[1mm] v-v_1 = \dfrac{1}{\xi}[c+\eta(u-u_1)] \end{array}\right\} \qquad (92)$$

$$\left.\begin{array}{l} v'^2+v_1'^2 = a'-u'^2-u_1'^2-2w'^2-\chi' \\[1mm] (v'+v_1')^2 = b'-(u'+u_1')^2-4w'^2 \\[1mm] v'-v_1' = \dfrac{1}{\xi'}[c'+\eta(u-u_1)] \end{array}\right\} \qquad (93)$$

In these equations, the u's and η''s are to be considered as given, since we expressed them in terms of given quantities. If one eliminates the v's by using equations (92), there remains only a single equation for ξ, which then also determines the v's. Likewise one can determine ξ', v' and v_1' from equations (93). If equation (88) is now satisfied, then we can express the variables $\xi, \eta, u, v \ldots$ separately in terms of the given quantities

$$a, b, c, \; w, a', b', c', w', A, B, C, A', B', C' \qquad (T)$$

The only equation relating these variables is (88). Equation (85) must therefore be satisfied for all values of the variables (T) which satisfy (88). Therefore ϕ must have the form $Ae^{-h\lambda}$. That w does not also appear in ϕ can easily be proved from the other collisions. Since already for the collisions considered, u_1 is completely arbitrary, and by considering all collisions naturally still more arbitrary quantities would be introduced, it does not seem likely that for some other force law other solutions would be possible. Yet I know of no other means of proof, at present, than to treat each force law by itself.

Inasmuch as we take it to be very probable that for the case of thermal equilibrium the function ϕ always has the form $Ae^{-h\lambda}$, we may now calculate E for any body for which thermal equilibrium has been established among its atoms. We use the expression (79) as a generally valid definition of E (ignoring constant quantities that may arise on account of the special nature of the problem).

If we wish to call E the entropy, we run up against the difficulty that the total entropy of two bodies would differ by a constant from the sum of the entropies of the individual bodies. We therefore prefer to consider the following expression, which differs from (79) only by a constant:

$$E^* = \iint \ldots f \log \left(\frac{f}{N}\right) dx_1 \ldots dw_r$$

Here N is the total number of molecules in the gas, while $fdx_1 \ldots dw_r$ is the number of those for which $x_1, y_1 \ldots w_r$ lie between the limits

$$x_1 \text{ and } x_1 + dx_1 \ldots w_r \text{ and } w_r + dw_r \qquad (S)$$

If we set

$$dx_1 dy_1 \ldots dz_r = d\sigma, \ du_1 dv_1 \ldots dw_r = ds, \ \frac{f}{N} = f^*$$

then f^* also has a simple meaning. $f^* ds d\sigma$ is the probability that a molecule has the state (S) (the time during which it has that state divided by the total time during which it moves).

We then have

$$E^* = N\iint f^* \log f^* \, ds d\sigma \tag{94}$$

For monatomic gases, if N is the total number of molecules in the gas, V the volume of the container, m the mass, and T the average kinetic energy of an atom, we have

$$f^* = \frac{1}{V\left(\dfrac{4\pi T}{3m}\right)^{\frac{3}{2}}} e^{-\frac{3m}{4T}(u^2 + v^2 + w^2)}$$

hence,

$$E^* = N\iint \ldots f^* \log f^* \, dx dy dz du dv dw$$

$$= -N \log\left[V\left(\frac{4\pi T}{3m}\right)^{\frac{3}{2}} \right] - \tfrac{3}{2} N$$

which, since m and N are constant, agrees up to a constant factor and constant additive term with the expression for the entropy of a monatomic gas. For gases with r-atomic molecules, we have

$$f^* = Ae^{-h(\chi + \Sigma mc^2/2)}$$

where χ is the potential; $\Sigma mc^2/2$ is the total kinetic energy of a molecule. Since $\iint f^* \, ds d\sigma = 1$,

$$T = \frac{3}{2h}, \quad A = \frac{1}{\left(\dfrac{2\pi}{mh}\right)^{\frac{3}{2}} \int e^{-h\chi} d\sigma}$$

One therefore finds that

$$E^* = N \log A - hN \frac{\int \chi e^{-h\chi} d\sigma}{\int e^{-h\chi} d\sigma} - \tfrac{3}{2} rN \tag{95}$$

In order to find the relation of the quantity E^* to the second law of thermodynamics in the form $\int dQ/T < 0$, we shall interpret the system of r mass points not, as previously, as a gas molecule, but rather as an entire body. (We shall call it the system A.) During a certain period of time it interacts with a second system (B), i.e. with a second body. The two bodies may have the same or different properties. Theoretically the effect of the interaction should depend not only on the nature of the force between A and B but also on the phases of both bodies at the time when the interaction begins. However, experience shows that this is not noticeable, doubtless because the effect of the phase is masked by the effect of the large number of molecules that are interacting. (A similar opinion has already been expressed recently by Clausius.†) In order to eliminate the effects of the phase, we shall replace the single system (A) by a large number (N) of equivalent systems distributed over many different phases, but which do not interact with each other. Let $f(t, x_1 \ldots w_r)dsd\sigma$ again be the number of systems with state (S), and set f/N equal to f^*. Similarly we assume that there are many systems of type (B). Their distribution is determined by a function $f'(t, x_1' \ldots w_r')$ similar to f. The functions f^* and f' may also be discontinuous, so that they have large values when the variables are very close to certain values determined by one or more equations, and otherwise vanishingly small. We may choose these equations to be those that characterize visible external motion of the body and the kinetic energy contained in it. In this connection it should be noted that the kinetic energy of visible motion corresponds to such a large deviation from the final equilibrium distribution of kinetic energy that it leads to an infinity in E^*, so that from the point of view of the second law of thermodynamics it acts like heat supplied from an infinite temperature. One of the (B) systems will now act on each of the (A) systems, and thus the beginning of the interaction will coincide with all different phases. All effects that do not depend on phase will then appear just as if only one (A) system acted on one (B) system in an arbitrary phase, and we know that thermal phenomena do not in fact depend on phase. The function f can therefore be chosen arbitrarily, insofar as it is not restricted by the conditions of total kinetic energy or visible motion of the

---

† [R. Clausius, *Ann. Phys.* **142**, 433 (1871); *Phil. Mag.* **42**, 161 (1871).]

body. The probability that an (A) system in state (S) interacts with a (B) system whose state is given by a similar condition is given by a formula similar to (52). From this it can be proved that E^* can only decrease. After a long-continued interaction (to establish temperature equilibrium) E^* attains its minimum, which occurs in general when $ff' = FF'$. If the bodies are at rest, then the solution of this equation is

$$f^* = Ae^{-h(\chi + \Sigma mc^2/2)}$$

where $f^*dsd\sigma$ is the probability that an (A) system has state (S). The quantity E, which is proportional to the entropy of all N (A) systems, is again given by equation (95). The entropy of a single (A) system is therefore $1/N$ of this, and is therefore proportional to

$$E^* = \iint f^* \log f^* \, dsd\sigma = \log A - h\frac{\int \chi e^{-h\chi}d\sigma}{\int e^{-h\chi}d\sigma} - \frac{3r}{2} \tag{96}$$

which agrees (up to a constant factor and addend) with the expression which I found in my previous paper.†

---

† L. Boltzmann, *Wien. Ber.* **63**, 712 (1871), Eqn. 18.

3

The Kinetic Theory of the Dissipation of Energy *

WILLIAM THOMSON

SUMMARY

The equations of motion in abstract dynamics are perfectly reversible; any solution of these equations remains valid when the time variable t is replaced by $-t$. Physical processes, on the other hand, are irreversible: for example, the friction of solids, conduction of heat, and diffusion. Nevertheless, the principle of dissipation of energy is compatible with a molecular theory in which each particle is subject to the laws of abstract dynamics.

Dissipation of energy, such as that due to heat conduction in a gas, might be entirely prevented by a suitable arrangement of Maxwell demons, operating in conformity with the conservation of energy and momentum. If no demons are present, the average result of the free motions of molecules will be to equalize temperature-differences. If we allowed this equalization to proceed for a certain time, and then reversed the motions of all the molecules, we would observe a disequalization. However, if the number of molecules is very large, as it is in a gas, any slight deviation from absolute precision in the reversal will greatly shorten the time during which disequalization occurs. In other words, the probability of occurrence of a distribution of velocities which will lead to disequalization of temperature for any perceptible length of time is very small. Furthermore, if we take account of the fact that no physical system can be completely isolated from its surroundings but is in principle interacting with all other molecules in the universe, and if we believe that the number of these latter molecules is infinite, then we may conclude that it is impossible for temperature-differences to arise spontaneously. A numerical calculation is given to illustrate this conclusion.

* Originally published in the *Proceedings of the Royal Society of Edinburgh*, Vol. 8, pp. 325–34 (1874); reprinted in Thomson's *Mathematical and Physical Papers*, Cambridge University Press, 1911, Vol. V, pp. 11–20.]

In abstract dynamics the instantaneous reversal of the motion of every moving particle of a system causes the system to move backwards, each particle of it along its old path, and at the same speed as before, when again in the same position. That is to say, in mathematical language, any solution remains a solution when t is changed into $-t$. In physical dynamics this simple and perfect reversibility fails, on account of forces depending on friction of solids; imperfect fluidity of fluids; imperfect elasticity of solids; inequalities of temperature, and consequent conduction of heat produced by stresses in solids and fluids; imperfect magnetic retentiveness; residual electric polarization of dielectrics; generation of heat by electric currents induced by motion; diffusion of fluids, solution of solids in fluids, and other chemical changes; and absorption of radiant heat and light. Consideration of these agencies in connection with the all-pervading law of the conservation of energy proved for them by Joule, led me twenty-three years ago to the theory of the dissipation of energy, which I communicated first to the Royal Society of Edinburgh in 1852, in a paper entitled " On a Universal Tendency in Nature to the Dissipation of Mechanical Energy."

The essence of Joule's discovery is the subjection of physical phenomena to dynamical law. If, then, the motion of every particle of matter in the universe were precisely reversed at any instant, the course of nature would be simply reversed for ever after. The bursting bubble of foam at the foot of a waterfall would reunite and descend into the water; the thermal motions would reconcentrate their energy, and throw the mass up the fall in drops re-forming into a close column of ascending water. Heat which had been generated by the friction of solids and dissipated by conduction, and radiation with absorption, would come again to the place of contact, and throw the moving body back against the force to which it had previously yielded. Boulders would recover from the mud the materials required to rebuild them into their previous jagged forms, and would become reunited to the mountain peak from which they had formerly broken away. And if also the materialistic hypothesis of life were true, living creatures would grow backwards, with conscious knowledge of the future, but no memory of the past, and would become again unborn. But the real phenomena of life infinitely transcend human science; and speculation regarding consequences of their imagined reversal is utterly unprofitable. Far otherwise, however, is it in respect to the reversal of the motions of matter uninfluenced by life, a very elementary consideration of which leads to the full explanation of the theory of dissipation of energy.

To take one of the simplest cases of the dissipation of energy, the conduction of heat through a solid—consider a bar of metal warmer at one end than the other, and left to itself. To avoid all needless complication of taking loss or gain of heat into account, imagine the bar to be varnished with a substance impermeable to heat. For the sake of definiteness, imagine the bar to be first given with one-half of it at one uniform temperature, and the other half of it at another uniform temperature. Instantly a diffusion of heat commences, and the distribution of temperature becomes continuously less and less unequal, tending to perfect uniformity, but never in any finite time attaining perfectly to this ultimate condition. This process of diffusion could be perfectly prevented by an army of Maxwell's " intelligent demons,"† stationed at the surface, or interface as we may call it with Professor James Thomson, separating the hot from the cold part of the bar. To see precisely how this is to be done, consider rather a gas than a solid, because we have much knowledge regarding the molecular motions of a gas, and little or no knowledge of the molecular motions of a solid. Take a jar with the lower half occupied by cold air or gas, and the upper half occupied with the air or gas of the same kind, but at a higher temperature; and let the mouth of the jar be closed by an air-tight lid. If the containing vessel were perfectly impermeable to heat, the diffusion

---

† The definition of a demon, according to the use of this word by Maxwell, is an intelligent being endowed with free-will and fine enough tactile and perceptive organization to give him the faculty of observing and influencing individual molecules of matter.

of heat would follow the same law in the gas as in the solid, though in the gas the diffusion of heat takes place chiefly by the diffusion of molecules, each taking its energy with it, and only to a small proportion of its whole amount by the interchange of energy between molecule and molecule; whereas in the solid there is little or no diffusion of substance, and the diffusion of heat takes place entirely, or almost entirely, through the communication of energy from one molecule to another. Fourier's exquisite mathematical analysis expresses perfectly the statistics of the process of diffusion in each case, whether it be " conduction of heat," as Fourier and his followers have called it, or the diffusion of substance in fluid masses (gaseous or liquid), which Fick showed to be subject to Fourier's formulas. Now, suppose the weapon of the ideal army to be a club, or, as it were, a molecular cricket bat; and suppose, for convenience, the mass of each demon with his weapon to be several times greater than that of a molecule. Every time he strikes a molecule he is to send it away with the same energy as it had immediately before. Each demon is to keep as nearly as possible to a certain station, making only such excursions from it as the execution of his orders requires. He is to experience no forces except such as result from collisions with molecules, and mutual forces between parts of his own mass, including his weapon. Thus his voluntary movements cannot influence the position of his centre of gravity, otherwise than by producing collision with molecules.

The whole interface between hot and cold is to be divided into small areas, each allotted to a single demon. The duty of each demon is to guard his allotment, turning molecules back, or allowing them to pass through from either side, according to certain definite orders. First, let the orders be to allow no molecules to pass from either side. The effect will be the same as if the interface were stopped by a barrier impermeable to matter and to heat. The pressure of the gas being by hypothesis equal in the hot and cold parts, the resultant momentum taken by each demon from any considerable number of molecules will be zero; and therefore he may so time his strokes that he shall never move to any considerable distance from his station. Now, instead of stopping and turning all the molecules from crossing his allotted area, let each demon permit a hundred molecules chosen arbitrary to cross it from the hot side; and the same number of molecules, chosen so as to have the same entire amount of energy and the same resultant momentum, to cross the other way from the cold side. Let this be done over and over again within certain small equal consecutive intervals of time, with care that if the specified balance of energy and momentum is not exactly fulfilled in respect to each successive hundred molecules crossing each way, the error will be carried forward, and as nearly as may be corrected, in respect to the next hundred. Thus, a certain perfectly regular diffusion of the gas both ways across the interface goes on, while the original different temperatures on the two sides of the interface are maintained without change.

Suppose, now, that in the original condition the temperature and pressure of the gas are each equal throughout the vessel, and let it be required to disequalize the temperature, but to leave the pressure the same in any two portions A and B of the whole space. Station the army on the interface as previously described. Let the orders now be that each demon is to stop all molecules from crossing his area in either direction except 100 coming from A, arbitrarily chosen to be let pass into B, and a greater number, having among them less energy but equal momentum, to cross from B to A. Let this be repeated over and over again. The temperature in A will be continually diminished and the number of molecules in it continually increased, until there are not in B enough of molecules with small enough velocities to fulfil the condition with reference to permission to pass from B to A. If after that no molecule be allowed to pass the interface in either direction, the final condition will be very great condensation and very low temperature in A; rarefaction and very high temperature in B; and equal pressures in A and B. The process of disequalization of temperature and density might be stopped at any time by changing the orders to those previously specified, and so permitting a certain degree of diffusion each way across the interface while maintaining a certain uniform difference of temperatures with equality of pressure on the two sides.

If no selective influence, such as that of the ideal " demon," guides individual molecules, the average result of their free motions and collisions must be to equalize the distribution of energy among them in the gross; and after a sufficiently long time, from the supposed initial arrangement, the difference of energy in any two equal volumes, each containing a very great number of molecules, must bear a very small proportion to the whole amount in either; or, more strictly speaking, the probability of the difference of energy exceeding any stated finite proportion of the whole energy in either is very small. Suppose now the temperature to have become thus very approximately equalized at a certain time from the beginning, and let the motion of every particle become instantaneously reversed. Each molecule will retrace its former path, and at the end of a second interval of time, equal to the former, every molecule will be in the same position, and moving with the same velocity, as at the beginning; so that the given initial unequal distribution of temperature will again be found, with only the difference that each particle is moving in the direction reverse to that of its initial motion. This difference will not prevent an instantaneous subsequent commencement of equalization, which, with entirely different paths for the individual molecules, will go on in the average according to the same law as that which took place immediately after the system was first left to itself.

By merely looking on crowds of molecules, and reckoning their energy in the gross, we could not discover that in the very special case we have just considered the progress was towards a succession of states, in which the distribution of energy deviates more and more from uniformity up to a certain time. The number of molecules being finite, it is clear that small finite deviations from absolute precision in the reversal we have supposed would not obviate the resulting disequalization of the distribution of energy. But the greater the number of molecules, the shorter will be the time during which the disequalizing will continue; and it is only

G

when we regard the number of molecules as practically infinite that we can regard spontaneous disequalization as practically impossible. And, in point of fact, if any finite number of perfectly elastic molecules, however great, be given in motion in the interior of a perfectly rigid vessel, and be left for a sufficiently long time undisturbed except by mutual impact and collisions against the sides of the containing vessel, it must happen over and over again that (for example) something more than $\frac{9}{10}$ths of the whole energy shall be in one-half of the vessel, and less than $\frac{1}{10}$th of the whole energy in the other half. But if the number of molecules be very great, this will happen enormously less frequently than that something more than $\frac{6}{10}$ths shall be in one-half, and something less than $\frac{4}{10}$ths in the other. Taking as unit of time the average interval of free motion between consecutive collisions, it is easily seen that the probability of these being something more than any stated percentage of excess above the half of the energy in one-half of the vessel during the unit of time from a stated instant, is smaller the greater the dimensions of the vessel and the greater the stated percentage. It is a strange but nevertheless a true conception of the old well-known law of the conduction of heat, to say that it is very improbable that in the course of 1000 years one-half of the bar of iron shall of itself become warmer by a degree than the other half; and that the probability of this happening before 1,000,000 years pass is 1000 times as great as that it will happen in the course of 1000 years, and that it certainly will happen in the course of some very long time. But let it be remembered that we have supposed the bar to be covered with an impermeable varnish. Do away with this impossible ideal, and believe the number of molecules in the universe to be infinite; then we may say one-half of the bar will never become warmer than the other, except by the agency of external sources of heat or cold. This one instance suffices to explain the philosophy of the foundation on which the theory of the dissipation of energy rests.

Take, however, another case, in which the probability may be readily calculated. Let an hermetically sealed glass jar of air contain 2,000,000,000,000 molecules of oxygen, and 8,000,000,000,000 molecules of nitrogen. If examined any time in the infinitely distant future, what is the number of chances against one that all the molecules of oxygen and none of nitrogen shall be found in one stated part of the vessel equal in volume to $\frac{1}{5}$th of the whole? The number expressing the answer in the Arabic notation has about 2,173,220,000,000 of places of whole numbers. On the other hand, the chance against there being exactly $\frac{2}{10}$ths of the whole number of particles of nitrogen, and at the same time exactly $\frac{2}{10}$ths of the whole number of particles of oxygen in the first specified part of the vessel, is only 4021×10^9 to 1.

Appendix

Calculation of probability respecting Diffusion of Gases

For simplicity, I suppose the sphere of action of each molecule to be infinitely small in comparison with its average distance from its nearest neighbour; thus, the sum of the volumes of the spheres of action of all the molecules will be infinitely small in proportion to the whole volume of the containing vessel. For brevity, space external to the sphere of action of every molecule will be called free space: and a molecule will be said to be in free space at any time when its sphere of action is wholly in free space; that is to say, when its sphere of action does not overlap the sphere of action of any other molecule. Let A, B denote any two particular portions of the whole containing vessel, and let a, b be the volumes of those portions. The chance that at any instant one individual molecule of whichever gas shall be in A is $a/(a+b)$, however many or few other molecules there may be in A at the same time; because its chances of being in any specified portions of free space are proportional to their volumes; and according to our supposition, even if all the other molecules were in A, the volume of free space in it would not be sensibly diminished by their presence. The chance that of n molecules in the whole space there shall be i stated individuals in A, and that the other $n-i$ molecules shall be at the same time in B, is

$$\left(\frac{a}{a+b}\right)^i \left(\frac{b}{a+b}\right)^{n-i}, \text{ or } \frac{a^i b^{n-i}}{(a+b)^n}$$

Hence the probability of the number of molecules in A being exactly i, and in B exactly $n-i$, irrespectively of individuals, is a fraction having for denominator $(a+b)^n$, and for numerator the term involving $a^i b^{n-i}$ in the expansion of this binomial; that is to say, it is

$$\frac{n(n-1)\ldots(n-i+1)}{1.2\ldots i}\left(\frac{a}{a+b}\right)^i \left(\frac{b}{a+b}\right)^{n-i}$$

If we call this T_i, we have

$$T_{i+1} = \frac{n-i}{i+1}\frac{a}{b}\, T_i$$

Hence T_i is the greatest term, if i is the smallest integer which makes

$$\frac{n-i}{i+1} < \frac{b}{a}$$

this is to say, if i is the smallest integer which exceeds

$$n\frac{a}{a+b} - \frac{b}{a+b}$$

Hence if a and b are commensurable, the greatest term is that for which

$$i = n\frac{a}{a+b}$$

To apply these results to the cases considered in the preceding article, put in the first place

$$n = 2 \times 10^{12}$$

this being the number of particles of oxygen; and let $i = n$. Thus, for the probability that all the particles of oxygen shall be in A, we find

$$\left(\frac{a}{a+b}\right)^{2 \times 10^{12}}$$

Similarly, for the probability that all the particles of nitrogen are in the space B, we find

$$\left(\frac{b}{a+b}\right)^{8 \times 10^{12}}$$

Hence the probability that all the oxygen is in A and all the nitrogen in B is

$$\left(\frac{a}{a+b}\right)^{2 \times 10^{12}} \times \left(\frac{b}{a+b}\right)^{8 \times 10^{12}}$$

Now by hypothesis

$$\frac{a}{a+b} = \frac{2}{10}$$

and therefore

$$\frac{b}{a+b} = \frac{8}{10}$$

hence the required probability is

$$\frac{2^{26 \times 10^{12}}}{10^{10^{18}}}$$

Call this $1/N$, and let log denote common logarithm. We have

$$\log N = 10^{13} - 26 \times 10^{12} \times \log 2 = (10 - 26 \log 2)$$
$$\times 10^{12} = 2173220 \times 10^6$$

This is equivalent to the result stated in the text above. The logarithm of so great a number, unless given to more than thirteen significant places, cannot indicate more than the number of places of the whole numbers in the answer to the proposed question, expressed according to the Arabic notation.

The calculation of T_i, when i and $n-i$ are very large numbers, is practicable by Stirling's theorem, according to which we have approximately

$$1.2 \ldots i = i^{i+\frac{1}{2}} \varepsilon^{-i} \sqrt{2\pi}$$

and therefore

$$\frac{n(n-1) \ldots (n-i+1)}{1.2 \ldots \ldots i} = \frac{n^{n+\frac{1}{2}}}{\sqrt{2\pi} i^{i+\frac{1}{2}} (n-i)^{n-i+\frac{1}{2}}}$$

Hence for the case

$$i = n \frac{a}{a+b}$$

which, according to the preceding formulae, gives T_i its greatest value, we have

$$T_i = \frac{1}{\sqrt{2\pi n e f}}$$

where

$$e = \frac{a}{a+b} \text{ and } f = \frac{b}{a+b}$$

Thus, for example, let $n = 2 \times 10^{12}$,

$$e = \cdot2, \quad f = \cdot8$$

we have

$$T_i = \frac{1}{800000\sqrt{\pi}} = \frac{1}{1418000}$$

This expresses the chance of there being 4×10^{11} molecules of oxygen in A, and 16×10^{11} in B. Just half this fraction expresses the probability that the molecules of nitrogen are distributed in exactly the same proportion between A and B, because the number of molecules of nitrogen is four times greater than of oxygen.

If n denote the number of the molecules of one gas, and n' that of the molecules of another, the probability that each shall be distributed between A and B in the exact proportion of the volume is

$$\frac{1}{2\pi e f \sqrt{nn'}}$$

The value for the supposed case of oxygen and nitrogen is

$$\frac{1}{2\pi \times \cdot16 \times 4 \times 10^{12}} = \frac{1}{4021 \times 10^{9}}$$

which is the result stated at the conclusion of the text above.

4

On the Relation of a General Mechanical Theorem to the Second Law of Thermodynamics *

LUDWIG BOLTZMANN

SUMMARY

Loschmidt has pointed out that according to the laws of mechanics, a system of particles interacting with any force law, which has gone through a sequence of states starting from some specified initial conditions, will go through the same sequence in reverse and return to its initial state if one reverses the velocities of all the particles. This fact seems to cast doubt on the possibility of giving a purely mechanical proof of the second law of thermodynamics, which asserts that for any such sequence of states the entropy must always increase.

Since the entropy would decrease as the system goes through this sequence in reverse, we see that the fact that entropy actually increases in all physical processes in our own world cannot be deduced solely from the nature of the forces acting between the particles, but must be a consequence of the initial conditions. Nevertheless, we do not have to assume a special type of initial condition in order to give a mechanical proof of the second law, if we are willing to accept a statistical viewpoint. While any individual non-uniform state (corresponding to low entropy) has the same probability as any individual uniform state (corresponding to high entropy), there are many more uniform states than non-uniform states. Consequently, if the initial state is chosen at random, the system is almost certain to evolve into a uniform state, and entropy is almost certain to increase.

* Originally published under the title: " Über die Beziehung eines allgemeine mechanischen Satzes zum zweiten Hauptsatze der Warmetheorie ", *Sitzungsberichte Akad. Wiss.*, Vienna, part II, **75**, 67–73 (1877); reprinted in Boltzmann's *Wissenschaftliche Abhandlungen*, Vol. 2, Leipzig, J. A. Barth, 1909, pp. 116–22.]

In his memoir on the state of thermal equilibrium of a system of bodies with regard to gravity, Loschmidt has stated a theorem that casts doubt on the possibility of a purely mechanical proof of the second law. Since it seems to me to be quite ingenious and of great significance for the correct understanding of the second law, yet in the cited memoir it has appeared in a more philosophical garb, so that many physicists will find it rather difficult to understand, I will try to restate it here.

If we wish to give a purely mechanical proof that all natural processes take place in such a way that

$$\int \frac{dQ}{T} \leqq 0$$

then we must assume the body to be an aggregate of material points. We take the force acting between these points to be a function of the relative positions of the points. When this force is known as a function of these relative positions, we shall say that the law of action of the force is known. In order to calculate the actual motion of the points, and therefore the state variations of the body, we must know also the initial positions and initial velocities of all the points. We say that the initial conditions must be given. If one tries to prove the second law mechanically, he always tries to deduce it from the nature of the law of action of the force without reference to the initial conditions, which are unknown. One therefore seeks to prove that—whatever may be the initial conditions—the state variations of the body will always take place in such a way that

$$\int \frac{dQ}{T} \leqq 0$$

We now assume that we are given a certain body as an aggregate of certain material points. The initial conditions at time zero shall be such that the body undergoes state variations for which

$$\int \frac{dQ}{T} \leqq 0$$

G•

We shall show that then, without changing the law of force, other initial conditions can be found for which conversely

$$\int \frac{dQ}{T} \geqq 0$$

Consider the positions and velocities of all the points after an arbitrary time t_1 has elapsed. We now take, in place of the original initial conditions, the following: all the material points† shall have the same initial positions at time zero that they had after time t_1 with the original initial conditions, and the same velocities but in the opposite directions. For brevity we shall call this state the one opposite to that previously found at time t_1.

It is clear that the points will pass through the same states as before but in the reverse order. The initial state which they had previously had at time zero, will now be reached after time t_1 has elapsed. Whereas previously we found

$$\int \frac{dQ}{T} \leqq 0$$

this quantity is now $\geqq 0$. The sign of this integral therefore does not depend on the force law but rather only on the initial conditions.‡ The fact that this integral is actually $\leqq 0$ for all processes in the world in which we live (as experience shows) is not due to the nature of the forces, but rather to the initial conditions. If, at time zero, the state of all material points in the universe were just the opposite of that which actually occurs at a much later time t_1, then the course of all events between times t_1 and zero would be reversed, so that

$$\int \frac{dQ}{T} \geqq 0$$

---

† By this we mean all the points of all bodies interacting with the one considered, either directly or indirectly. Strictly speaking one has to include all the points in the universe, since a complex of bodies that does not interact at all with the other bodies in the universe cannot actually be found, even though we can imagine it.

‡ It need not be mentioned that if the forces act in such a way that this is not true, for example if they are dynamical, then the following also loses its applicability.

Thus any attempt to prove from the nature of bodies and of the
the force law, without taking account of initial conditions, that

$$\int \frac{dQ}{T} \leqq 0$$

must necessarily be futile. One sees that this conclusion has
great seductiveness and that one must call it an interesting
sophism. In order to locate the source of the fallacy in this
argument, we shall imagine a system of a finite number of material
points which does not interact with the rest of the universe.

We imagine a large but not infinite number of absolutely elastic
spheres, which move in a closed container whose walls are completely rigid and likewise absolutely elastic. No external forces
act on our spheres. Suppose that at time zero the distribution of
spheres in the container is not uniform; for example, suppose that
the density of spheres is greater on the right than on the left, and
that the ones in the upper part move faster than those in the lower,
and so forth. The sophism now consists in saying that, without
reference to the initial conditions, it cannot be proved that the
spheres will become uniformly mixed in the course of time. For
the initial conditions which we originally assumed, the spheres will
be almost always uniform at time t_1, for example. We can then
choose in place of the original initial conditions the distribution of
states which is just the opposite of the one which would occur (in
consequence of the original initial conditions) after time t_1 has
elapsed. Then the spheres would sort themselves out as time
progresses, and at time t_1 they would acquire a completely non-
uniform distribution of states, even though the initial distribution
of states was almost uniform.

We must make the following remark: a proof, that after a
certain time t_1 the spheres must necessarily be mixed uniformly,
whatever may be the initial distribution of states, cannot be given.
This is in fact a consequence of probability theory, for any non-
uniform distribution of states, no matter how improbable it may
be, is still not absolutely impossible. Indeed it is clear that any
individual uniform distribution, which might arise after a certain

time from some particular initial state, is just as improbable as an individual non-uniform distribution; just as in the game of Lotto, any individual set of five numbers is as improbable as the set 1, 2, 3, 4, 5. It is only because there are many more uniform distributions than non-uniform ones that the distribution of states will become uniform in the course of time. One therefore cannot prove that, whatever may be the positions and velocities of the spheres at the beginning, the distribution must become uniform after a long time; rather one can only prove that infinitely many more initial states will lead to a uniform one after a definite length of time than to a non-uniform one. Loschmidt's theorem tells us only about initial states which actually lead to a very non-uniform distribution of states after a certain time t_1; but it does not prove that there are not infinitely many more initial conditions that will lead to a uniform distribution after the same time. On the contrary, it follows from the theorem itself that, since there are infinitely many more uniform than non-uniform distributions, the number of states which lead to uniform distributions after a certain time t_1 is much greater than the number that leads to non-uniform ones, and the latter are the ones that must be chosen, according to Loschmidt, in order to obtain a non-uniform distribution at t_1.

One could even calculate, from the relative numbers of the different state distributions, their probabilities, which might lead to an interesting method for the calculation of thermal equilibrium.† In just the same way one can treat the second law. It is only in some special cases that it can be proved that, when a system goes over from a non-uniform to a uniform distribution of states, then $\int dQ/T$ will be negative, whereas it is positive in the opposite case. Since there are infinitely many more uniform than non-uniform distributions of states, the latter case is extraordinarily improbable and can be considered impossible for practical purposes; just as it may be considered impossible that if one starts with oxygen and nitrogen mixed in a container, after a month one will find chemi-

---

† [Following up this remark, Boltzmann developed soon afterward his statistical method for calculating equilibrium properties, based on the relation between entropy and probability: *Wien. Ber.* **76**, 373 (1877).]

cally pure oxygen in the lower half and nitrogen in the upper half, although according to probability theory this is merely very improbable but not impossible.

Nevertheless Loschmidt's theorem seems to me to be of the greatest importance, since it shows how intimately connected are the second law and probability theory, whereas the first law is independent of it. In all cases where $\int dQ/T$ can be negative, there is also an individual very improbable initial condition for which it may be positive; and the proof that it is almost always positive can only be carried out by means of probability theory. It seems to me that for closed paths of the atom, $\int dQ/T$ must always be zero, which can therefore be proved independently of probability theory. For unclosed paths it can also be negative. I will mention here a peculiar consequence of Loschmidt's theorem, namely that when we follow the state of the world into the infinitely distant past, we are actually just as correct in taking it to be very probable that we would reach a state in which all temperature differences have disappeared, as we would be in following the state of the world into the distant future. This would be similar to the following case: if we know that in a gas at a certain time there is a non-uniform distribution of states, and that the gas has been in the same container without external disturbance for a very long time, then we must conclude that much earlier the distribution of states was uniform and that the rare case occurred that it gradually became non-uniform. In other words: any non-uniform distribution evolves into an almost uniform one after a long time t_1. The one opposite to this latter one evolves, after the same time t_1, into the initial non-uniform one (more precisely, into the opposite of it). The distribution opposite to the initial one would however, if chosen as an initial distribution, likewise evolve into a uniform distribution after time t_1.

If perhaps this reduction of the second law to the realm of probability makes its application to the entire universe appear dubious, yet the laws of probability theory are confirmed by all experiments carried out in the laboratory.

5

On the Three-body Problem and the Equations of Dynamics *

HENRI POINCARÉ

SUMMARY

Poisson attempted to show that a mechanical system is stable, in the sense that it will eventually return to a configuration very close to its initial one. He did not think that all solutions of the equations of dynamics would be stable: stability may depend on the initial conditions.

It is proved that there are infinitely many ways of choosing the initial conditions such that the system will return infinitely many times as close as one wishes to its initial position. There are also an infinite number of solutions that do not have this property, but it is shown that these unstable solutions can be regarded as " exceptional " and may be said to have zero probability.

§ 1. Notations and definitions

We consider a system of differential equations

$$\frac{dx_1}{dt} = X_1, \ \frac{dx_2}{dt} = X_2, \ldots, \ \frac{dx_n}{dt} = X_n \tag{1}$$

where t represents the independent variable which we shall call the time, $x_1, x_2 \ldots, x_n$ are the unknown functions, and $X_1, X_2, \ldots, X_n$

---

* [Originally published under the title: " Sur le problème des trois corps et les équations de dynamique ", *Acta Mathematica* **13**, pp. 1–270 (1890); reprinted in the *Oeuvres de Henri Poincaré*, Gauthier-Villars, Paris, 1952, **7**, pp. 262–479. The extracts given here are translated from pp. 8, 10 and 67–73 of the original; the definition of integral invariants, summarized in an editor's note, is taken from pp. 52–5.]

are given functions of $x_1, x_2, \ldots, x_n$. We assume in general that the functions $X_1, X_2, \ldots, X_n$ are analytic and uniform for all real values of the $x_1, x_2, \ldots, x_n$.

.   .   .   .

We consider in particular the case $n = 3$; we can then regard x_1, x_2 and x_3 as the coordinates of a point P in space. . . . When we vary the time t, the point P will describe a certain curve in space which we call a *trajectory*. To each particular solution of equations (1) there corresponds a trajectory, and conversely.

If the functions X_1, X_2 and X_3 are uniform, one and only one trajectory will pass through each point in space. The only exception occurs when one of these three functions becomes infinite, or when all three are zero.

.   .   .   .

[A relation of the form

$$F(x_1, x_2, \ldots, x_n) = \text{constant}$$

where the $x_1, x_2, \ldots, x_n$ are solutions of the system of equations (1), is called an *integral* of those equations. More generally we may have a relation such as

$$F_1(x_1, x_2, \ldots, x_n, x'_1, x'_2, \ldots, x'_n) = \text{constant}$$

where $x'_1, x'_2, \ldots, x'_n$ are another solution of the same system of equations; or we may have such a relation involving any number of such solutions. A relation which involves an integral over an infinite number of such solutions is known as an *integral invariant* of the system. The most important example is an integral over all the solutions contained in a certain volume of space:

$$\iiint dx\,dy\,dz = \text{constant}$$

In other words, the volume occupied by a specified set of initial coordinates, corresponding to a region of possible positions of the point P, remains invariant as each of these points moves along the trajectory determined by the system of equations. This "conservation of volume in phase space" is also known as Liouville's theorem in statistical mechanics.]

.   .   .   .

§ 8. Use of integral invariants

The interest of integral invariants results from the following theorems of which we shall make frequent use.

We have defined stability above [§1] by saying that the mobile point P should stay at a finite distance from its starting place; one sometimes understands stability in a different sense, however. In order to have stability, it is necessary that the point P should return after a sufficiently long time, if not to its initial position, at least to a position as close as one wishes to this initial position.

It is in this latter sense that Poisson† means stability. When he showed that, if one takes account of the second powers of the masses, the major axes remain invariant, he was only interested in showing that the series expansions of the major axes contain only periodic terms of the form $\sin \alpha t$ or $\cos \alpha t$, or mixed terms of the form $t \sin \alpha t$ or $t \cos \alpha t$, without containing any secular terms of the form t or t^2. This does not mean that the major axes cannot exceed a certain value, for a mixed term $t \cos \alpha t$ can increase beyond any limit; it means only that the major axes return infinitely many times to their original values.

Does stability, in the sense of Poisson, apply to all solutions? Poisson did not think so, for his proof assumed explicitly that the average motions are not commensurable; it therefore does not apply for any arbitrary initial conditions of motion.

The existence of asymptotic solutions, which we will establish later on, is a sufficient proof that if the initial position of the point P is appropriately chosen, the point P will not pass infinitely many times as close as one wishes to this initial position.

But I propose to establish that in the particular case of the three-body problem, one can choose the initial position of P in an infinite number of ways such that it does return infinitely many times as close as one wishes to its initial position.

---

† [S. D. Poisson, *Nouveau Bulletin des Sciences d la Société Philomatique de Paris* **1**, 191 (1808); *Mémoires de l'Academie Royale des Sciences de l'Institut de France* **7**, 199 (1827).]

In other words, there will be an infinity of particular solutions of the problem which do not possess stability in the second sense of the word, that of Poisson; but there will be an infinity that do. I will add that the former can be regarded as " exceptional "; I will attempt later on to clarify the exact meaning that I attach to this word.

Let $n = 3$ and imagine that x_1, x_2, x_3 represent the coordinates of a point P in space.

THEOREM I.   Suppose that the point P remains at a finite distance, and that the volume $\int dx_1 dx_2 dx_3$ is an integral invariant; if one considers any region r_0 whatever, no matter how small may be this region, there will be trajectories which traverse it infinitely many times.

In particular, suppose that P does not leave a bounded region R. I call V the volume of this region R.

Now let us imagine a very small region r_0, and call v the volume of this region. Through each point of r_0 there passes a trajectory which one may regard as being traversed by a mobile point following the law defined by our differential equations. Let us then consider an infinity of mobile points filling the region r_0 at time 0 and afterwards moving in accordance with this law. At time τ they fill a certain region r_1, at time 2τ a region r_2, etc., and at time $n\tau$ a region r_n. I may assume that τ is so large, and r_0 so small, that r_0 and r_1 have no point in common.

The volume being an integral invariant, the various regions $r_0, r_1, \ldots, r_n$ will have the same volume v. If these regions have no points in common, the total volume will be greater than nv; but since all these regions are inside R, the total volume is smaller than V. If one has

$$n > \frac{V}{v}$$

it is necessary that at least two of our regions must have a part in common. Let r_p and r_q be these two regions $(q > p)$. If r_p and r_q have a common point, it is clear that r_0 and r_{p-q} must have a common point.

More generally, if one cannot find k regions having a common part, then any point in space can belong to no more than $k-1$ of the regions $r_0, r_1, \ldots, r_n$. The total volume occupied by these regions will then be greater than $nv/(k-1)$. If then one has

$$n > (k-1)\frac{V}{v}$$

there must be k regions having a common part. Let

$$r_{p_1}, \; r_{p_2}, \ldots, \; r_{p_k}$$

be these regions. Then

$$r_0, \; r_{p_2-p_1}, \; r_{p_3-p_1}, \; \ldots, \; r_{p_k-p_1}$$

will have a common part.

We now consider the question from another viewpoint. By analogy with the terminology of the previous section we agree to say that the region r_n is the nth consequent of r_0 and r_0 is the nth antecedent of r_n.

Suppose then that r_p is the first of the successive consequents of r_0 which has a part in common with r_0. Let r_0' be this common part; let s_0' be the pth antecedent of r_0' which will also be part of r_0 since its pth consequent is part of r_p.

Finally let r_{p_1}' be the first of the consequents of r_0' which has a part in common with r_0'; let r_0'' be this common part; its p_1th antecedent will be part of r_0' and consequently of r_0, and its $p+p_1$ antecedent, which I will call s_0'', will be part of s_0' and therefore of r_0.

Thus s_0'' will be part of r_0 as well as of its pth and $(p+p_1)$th consequents.

Continuing in the same way, we can form r_0''' from r_0'' just as we formed r_0'' from r_0', and r_0' from r_0; so we likewise form $r_0^{IV}, \ldots, r_0^n, \ldots$

I will assume that the first of the successive consequents of r_0^n which has a part in common with r_0^n will be that of order p_n.

I will call s_0^n the antecedent of order $p+p_1+p_2+p_{n-1}$ of r_0^n.

The s_0^n will be part of r_0 as well as of its n consequents of order:

$$p, \; p+p_1, \; p+p_1+p_2, \ldots, \; p+p_1+p_2+ \ldots +p_{n-1}$$

Moreover, s_0^n will be part of s_0^{n-1}; s_0^{n-1} will be part of s_0^{n-2}, and so forth.

There will then be points that belong at the same time to regions $r_0, s_0', s_0'', \ldots, s_0^n, s_0^{n+1}, \ldots$ ad infinitum. The collection of these points will form a region σ which may be reduced to one or several points.

Then the region σ will be part of r_0 as well as its consequents of order $p, \; p+p_1, \ldots, p+p_1+ \ldots +p_n, \; p+p_1+ \ldots +p_n+p_{n+1}, \ldots$, ad infinitum.

In other words, all trajectories starting from one of the points of σ will go through the region r_0 infinitely many times.

Q. E. D.

COROLLARY. It follows from the preceding that there exists an infinite number of trajectories which can pass through the region r_0 infinitely many times; but there may also exist others that traverse it only a finite number of times. I now propose to explain why these latter trajectories may be considered " exceptional ".

This expression not having by itself a precise sense, I am first obliged to complete its definition.

We shall agree to say that the probability that the initial position of the mobile point P belongs to a certain region r_0 is to the probability that this initial position belongs to another region r_0' in the same ratio as the volume of r_0 to the volume of r_0'.

Probabilities being defined thus, I propose to establish that the probability that a trajectory starting from a point of r_0 does not go through this region more than k times is zero, however large k may be, and however small the region r_0. This is what I mean when I say that the trajectories that traverse r_0 only a finite number of times are exceptional.

I assume that the initial position of the point P belongs to r_0 and I propose to calculate the probability that the trajectory starting from this point does not pass through r_0 as many as $k+1$ times in the interval between time 0 and time $n\tau$.

We saw that if the volume v of r_0 is such that

$$n > \frac{kV}{v}$$

one can find $k+1$ regions which I call

$$r_0, r_{\alpha_1}, r_{\alpha_2}, \ldots, r_{\alpha_k}$$

and which will have a common part. Let s_{α_k} be this common part, let s_0 be its antecedent of order α_k; and designate by s_p the pth consequent of s_0.

I say that if the initial position of P belongs to s_0, the trajectory starting from this point traverses $k+1$ times at least the region r_0 between time 0 and time $n\tau$. In particular, the mobile point which describes this trajectory will be in the region s_0 at time 0, in s_p at $p\tau$, and in s_n at $n\tau$. It therefore must go through the following regions between 0 and $n\tau$:

$$s_0, s_{\alpha_k-\alpha_{k-1}}, s_{\alpha_k-\alpha_{k-2}}, \ldots, s_{\alpha_k-\alpha_2}, s_{\alpha_k-\alpha_1}, s_{\alpha_k}$$

Now I say that all these regions are part of r_0. In particular, s_{α_k} is part of r_0 by definition; s_0 is part of r_0 because its α_kth consequent s_{α_k} is part of r_{α_k}; and in general $s_{\alpha_k-\alpha_i}$ will be part of r_0 because its α_ith consequent s_{α_k} is part of r_{α_i}.

Then the mobile point traverses the region r_0 at least $k+1$ times.

Q. E. D.

Now let σ_0 be the part of r_0 which belongs neither to s_0 nor to any similar region, so that the trajectories starting from the various points of σ_0 do not go through r_0 at least $k+1$ times between times 0 and $n\tau$. Let w be the volume of σ_0.

The probability sought, that is to say the probability that our trajectory does not traverse r_0 as many as $k+1$ times during this time-interval, will then be w/v.

Now by hypothesis any trajectory starting from σ_0 does not traverse r_0 as many as $k+1$ times, and *a fortiori* does not traverse σ_0 as many as $k+1$ times. One then has

$$w < \frac{kV}{v}$$

and our probability will be smaller than

$$\frac{kV}{nv}$$

However large k may be, and however small v may be, one can always take n so large that this expression will be as small as one wishes.   Then the probability that our trajectory, which starts from a point of r_0, does not traverse this region more than k times between times 0 and ∞, is zero.

Q. E. D.

EXTENSION OF THEOREM I. We assumed that
1. $n = 3$.
2. The volume is an invariant integral.
3. The point P is required to remain at a finite distance.

The theorem is still true if the volume is not an invariant integral, provided there still exists any other positive invariant:

$$\int M dx_1 dx_2 dx_3$$

It is still true if $n > 3$, if there exists a positive invariant

$$\int M dx_1 dx_2 dx_3$$

and if $x_1, x_2, \ldots, x_n$, the coordinates of the point P in the n-dimensional space, are required to remain finite.

There is an even stronger statement which is valid.

Suppose that $x_1, x_2, \ldots, x_n$ are not required to remain finite, but the positive integral invariant

$$\int M dx_1 dx_2 \ldots dx_n$$

extended over the entire n-dimensional space has a finite value. The theorem will still be true then.

Here is a case that frequently arises.

Suppose that one knows an integral of equations (1):

$$F(x_1, x_2, \ldots, x_n) = \text{constant}$$

If $F = $ constant is the general equation of a system of closed surfaces in the n-dimensional space, if in other words F is a uniform function which becomes infinite whenever one of the variables $x_1, x_2, \ldots, x_n$ ceases to be finite, then it is clear that x_1, $x_2, \ldots x_n$ will always remain finite since F retains a finite constant value; one then has satisfied the conditions of the theorem.

But suppose that the surfaces $F = $ constant are not closed; it may happen nevertheless that the invariant positive integral

$$\int M dx_1 dx_2 \ldots dx_n$$

extended over the entire set of values of the x's such that

$$C_1 < F < C_2$$

has a finite value; the theorem will then still be true.

6

Mechanism and Experience *

HENRI POINCARÉ

SUMMARY

The advocates of the mechanistic conception of the universe have met with several obstacles in their attempts to reconcile mechanism with the facts of experience. In the mechanistic hypothesis, all phenomena must be reversible, while experience shows that many phenomena are irreversible. It has been suggested that the apparent irreversibility of natural phenomena is due merely to the fact that molecules are too small and too numerous for our gross senses to deal with them, although a " Maxwell demon " could do so and would thereby be able to prevent irreversibility.

The kinetic theory of gases is up to now the most serious attempt to reconcile mechanism and experience, but it is still faced with the difficulty that a mechanical system cannot tend toward a permanent final state but must always return eventually to a state very close to its initial state [Selection 5]. This difficulty can be overcome only if one is willing to assume that the universe does not tend irreversibly to a final state, as seems to be indicated by experience, but will eventually regenerate itself and reverse the second law of thermodynamics.

Everyone knows the mechanistic conception of the universe which has seduced so many good men, and the different forms in which it has been dressed.

Some represent the material world as being composed of atoms which move in straight lines because of their inertia; the velocity and direction of this motion cannot change except when two atoms collide.

Others allow action at a distance, and suppose that the atoms

---

* [Originally published under the title: " Le mécanisme et l'expérience ", *Revue de Metaphysique et de Morale* 1, pp. 534–7 (1893).]

exert on each other an attraction (or a repulsion) which depends on their distance according to some law.

The first viewpoint is clearly only a particular case of the second; what I am going to say will be as true of one as of the other. The most important conclusions apply also to Cartesian mechanism, in which one assumes a continuous matter.

It would perhaps be appropriate to discuss here the metaphysical difficulties that underlie these conceptions; but I do not have the necessary authority for that. Rather than discussing with the readers of this review that which they know better than I do, I prefer to speak of subjects with which they are less familiar, but which may interest them indirectly.

I am going to concern myself with the obstacles which the mechanists have encountered when they wished to reconcile their system with experimental facts, and the efforts which they have made to overcome or circumvent them.

In the mechanistic hypothesis, all phenomena must be *reversible*; for example, the stars might traverse their orbits in the retrograde sense without violating Newton's law; this would be true for any law of attraction whatever. This is therefore not a fact peculiar to astronomy; reversibility is a necessary consequence of all mechanistic hypotheses.

Experience provides on the contrary a number of irreversible phenomena. For example, if one puts together a warm and a cold body, the former will give up its heat to the latter; the opposite phenomenon never occurs. Not only will the cold body not return to the warm one the heat which it has taken away when it is in direct contact with it; no matter what artifice one may employ, using other intervening bodies, this restitution will be impossible, at least unless the gain thereby realized is compensated by an equivalent or large loss. In other words, if a system of bodies can pass from state A to state B by a certain path, it cannot return from B to A, either by the same path or by a different one. It is this circumstance that one describes by saying that not only is there not *direct reversibility*, but also there is not even *indirect reversibility*.

There have been many attempts to escape this contradiction; first there was Helmholtz's hypothesis of "hidden movements". Recall the experiment made by Foucault and Panthéon with a very long pendulum. This apparatus seems to turn slowly, indicating the rotation of the earth. An observer who does not know about the movement of the earth would certainly conclude that mechanical phenomena are irreversible. The pendulum always turns in the same sense, and there is no way to make it turn in the opposite sense; to do that it would be necessary to change the sense of rotation of the earth. Such a change is of course impractical, but for us it is conceivable; it would not be so for a man who believed our planet to be immobile.

Can one not imagine that there exist similar motions in the molecular world, which are hidden from us, which we have not taken account of, and of which we cannot change the sense?

This explanation is seductive, but it is insufficient; it shows why there is not *direct* reversibility; but one can show that it still requires *indirect* reversibility.

The English have proposed a completely different hypothesis. To explain it, I will make use of a comparison: if one had a hectolitre of wheat and a grain of barley, it would be easy to hide this grain in the middle of the wheat; but it would be almost impossible to find it again, so that the phenomenon appears to be in a sense irreversible. This is because the grains are small and numerous; the apparent irreversibility of natural phenomena is likewise due to the fact that the molecules are too small and too numerous for our gross senses to deal with them.

To clarify this explanation, Maxwell introduced the fiction of a " demon " whose eyes are sharp enough to distinguish the molecules, and whose hands are small and fast enough to grab them. For such a demon, if one believes the mechanists, there would be no difficulty in making heat pass from a cold to a warm body.

The development of this idea has given rise to the kinetic theory of gases, which is up to now the most serious attempt to reconcile mechanism and experience.

But all the difficulties have not been overcome.

A theorem, easy to prove, tells us that a bounded world, governed only by the laws of mechanics, will always pass through a state very close to its initial state. On the other hand, according to accepted experimental laws (if one attributes absolute validity to them, and if one is willing to press their consequences to the extreme), the universe tends toward a certain final state, from which it will never depart. In this final state, which will be a kind of death, all bodies will be at rest at the same temperature.

I do not know if it has been remarked that the English kinetic theories can extricate themselves from this contradiction. The world, according to them, tends at first toward a state where it remains for a long time without apparent change; and this is consistent with experience; but it does not remain that way forever, if the theorem cited above is not violated; it merely stays there for an enormously long time, a time which is longer the more numerous are the molecules. This state will not be the final death of the universe, but a sort of slumber, from which it will awake after millions of millions of centuries.

According to this theory, to see heat pass from a cold body to a warm one, it will not be necessary to have the acute vision, the intelligence, and the dexterity of Maxwell's demon; it will suffice to have a little patience.

One would like to be able to stop at this point and hope that some day the telescope will show us a world in the process of waking up, where the laws of thermodynamics are reversed.

Unfortunately, other contradictions arise; Maxwell made ingenious efforts to conquer them. But I am not sure that he succeeded. The problem is so complicated that it is impossible to treat it with complete rigour. One is then forced to make certain simplifying hypotheses; are they legitimate, are they self-consistent? I do not believe they are. I do not wish to discuss them here; but there is no need for a long discussion in order to challenge an argument of which the premises are apparently in contradiction with the conclusion, where one finds in effect reversibility in the premises and irreversibility in the conclusion.

Thus the difficulties that concern us have not been overcome, and it is possible that they never will be. This would amount to a definite condemnation of mechanism, if the experimental laws should prove to be distinctly different from the theoretical ones.

7

On a Theorem of Dynamics and the Mechanical Theory of Heat *

ERNST ZERMELO

SUMMARY

Poincaré's recurrence theorem [Selection 5] shows that irreversible processes are impossible in a mechanical system. A simple proof of this theorem is given.

The kinetic theory cannot provide an explanation of irreversible processes unless one makes the implausible assumption that only those initial states that evolve irreversibly are actually realized in nature, while the other states, which from a mathematical viewpoint are more probable, actually do not occur. It is concluded that it is necessary to formulate either the second law of thermodynamics or the mechanical theory of heat in an essentially different way, or else give up the latter theory altogether.

In the second chapter of Poincaré's prize essay on the three-body problem,† there is proved a theroem from which it follows that the usual description of the thermal motion of molecules, on which is based for example the kinetic theory of gases, requires an important modification in order that it be consistent with the thermodynamic law of increase of entropy. Poincaré's theorem says that *in a system of mass-points under the influence of forces that depend only on position in space, in general any state of motion (charac-*

---

* [Originally published under the title: " Uber einen Satz der Dynamik und die mechanische Warmetheorie ", *Annalen der Physik* **57**, pp. 485–94 (1896).]

† Poincaré, " Sur les équations de la dynamique et le problème des trois corps ", *Acta Mathematica* **13**, pp. 1–270 (1890); the theorem referred to is on pp. 67–72. [Selection 5]

terized by configurations and velocities) must recur arbitrarily often, at least to any arbitrary degree of approximation even if not exactly, provided that the coordinates and velocities cannot increase to infinity. Hence, in such a system *irreversible processes are impossible* since (aside from singular initial states) no single-valued continuous function of the state variables, such as entropy, can continually increase; if there is a finite increase, then there must be a corresponding decrease when the initial state recurs. Poincaré, in the essay cited, used his theorem for astronomical discussions on the stability of sun systems; he does not seem to have noticed its applicability to systems of molecules or atoms and thus to the mechanical theory of heat, although he has taken especial interest in fundamental questions of thermodynamics, and by another method he has tried to show that irreversible processes cannot always be explained by Helmholtz's theory of " monocyclic systems.†,‡ In order not to have to assume an acquaintance with Poincaré's long and—to many physicists—difficultly accessible work, I will sketch here the simplest possible proof of the theorem mentioned above.

Let N be the number of mass-points; there will be $n = 6N$ state quantities, i.e. the $3N$ coordinates and $3N$ velocity components, which we denote by $x_1, x_2, \ldots, x_n$. The time-derivatives of the coordinates will be identical with the corresponding velocities; the derivatives of the velocities, i.e. the accelerations, will be the forces, which according to our assumption are single-valued continuous functions of the coordinates. The former will therefore be independent of the coordinates, and the latter of the velocities, and the differential equations for the motion are of the form

$$\frac{dx_\mu}{dt} = X_\mu(x_1, x_2, \ldots x_n) \qquad (1)$$

$$(\mu = 1, 2, \ldots n)$$

† Poincaré, *Compt. rend. Acad. Sci., Paris* **108,** pp. 550–2 (1889); *Vorles. über Thermodynamik,* pp. 294–6.

‡ [H. v. Helmholtz, *Sitzber. K. Preuss. Akad. Wiss., Berlin,* **159,** 311, 755 (1884); *J. f. reine und angew. Math.* **97,** 111, 317 (1884).]

where none of the X_μ functions depends on the corresponding variable $x\mu$; thus we have the relations:

$$\frac{\partial X_1}{\partial x_1} + \frac{\partial X_2}{\partial x_2} + \ldots + \frac{\partial Xn}{\partial x_n} = 0 \qquad (2)$$

In such a system (1) of differential equations of the first order, there corresponds to an arbitrary initial state P_0:

$$x_1 = \xi_1, x_2 = \xi_2 \ldots x_n = \xi_n, (t = t_0)$$

a definite transformed state P at time t, expressed by the integral of (1):

$$x_\mu = \phi_\mu(t-t_0, \xi_1, \xi_2, \ldots \xi_n) \qquad (3)$$
$$(\mu = 1, 2, \ldots n)$$

where the ϕ_μ are singled-valued continuous functions of all their arguments, which, regardless of the choice of initial time t_0, are determined only by the functions X_μ. These relations are equally valid for previous or later times, i.e. for either negative or positive values of $t-t_0$; the initial state P_0 is an arbitrarily chosen phase of the motion, which does not have to precede any other phase. Likewise there corresponds to any *region* g_0 filled with initial states, describable by means of relations of the form:

$$F(\xi_1, \ldots \xi_n) < 0$$

a definite transformed region $g = g_t$ at time t. To these regions there correspond the n-fold integrals,

$$\gamma_0 = \int d\xi_1 d\xi_2 \ldots d\xi_n$$

which we shall call the " extension " of g_0, and in general another extension of g.

$$\gamma = \int dx_1 dx_2 \ldots dx_n$$

However, in the special case where the functions X_μ satisfy the condition (2), Liouville's theorem† says that the second integral

† Jacobi, *Dynamik*, p. 93; Kirchhoff, *Theorie der Wärme*, pp. 142–4.

is equal to the first, and indeed is independent of time, however one may choose the regions g_0 or g (either of which determines the other). Thus one can write:

$$dy = dx_1 dx_2 \ldots dx_n = d\gamma_0 = \text{constant} \qquad (4)$$

" The future states that correspond to the initial states in any region will fill at a given time a region of equal extension."

An arbitrary region g_0 of states therefore goes continuously into a new region $g = g_t$, the " phases " of its transformation, which always has the same extension γ. All these " *later* " phases g_t ($t > 0$) taken together form another continuous region G_0, the " *future* " of g_0, i.e. the aggregate of all states that arise from g_0 sometime in the future, in a finite time. This region $G = G_0$ will be finite and will have a finite extension $\Gamma \geq \gamma$, if we assume that the quantities $x_1, x_2 \ldots, x_n$ can never exceed certain finite limits, for all initial states in g_0. While the region g changes with time, at the same time all its " later phases " will change into the following ones, and G also changes so that it represents at any instant t the " future " of the corresponding phase g_t. According to the definition of future, this change must take place in such a way that while earlier states may leave G, no new ones can enter: each phase of G includes all the later ones in itself, so that the extension Γ may *only decrease*. But since according to (4) this extension must remain constant, the states that leave G cannot fill a region of finite extension, so that their number is vanishingly small compared to the remaining ones; we therefore call them *singular*. Now if g_0 is contained in G_0, then it is also completely or almost completely contained in each following phase G_τ, the future of g_τ, for an arbitrarily large time interval τ. But this means that there are always states in g_τ that will later transform to states in g_0, and, conversely, states in g_0 that after time τ will again return to g_0. These latter states are found in any part of the region, no matter how small, to which the same conditions apply as to g_0; and they will be connected together, since every recurrent state must be surrounded by a neighbourhood of states that also are recurrent; hence, these states fill the entire region g_0 with the exception of singular states of total extension zero. If one excludes all these singular states, then for *any* finite time τ there will be a region g', which no longer need be continuous but still contains the overwhelming majority of the states of g_0. These states of g' will always recur once after an arbitrary time, and therefore will recur infinitely often and come back arbitrarily close to their initial values if one chooses g_0 sufficiently small.

We have proved Poincaré's theorem in complete generality, although for our present purpose it suffices to show that the states in g_0 will return at least *once* to g_0. From this it follows directly that *there can be no single-valued continuous function $S = S(x_1, x_2 \ldots x_n)$ of the states that always increases for all initial states in some region, no matter how small the region.* For then S would have to increase from a value less than R to a value greater than R during time τ, for some initial state P_0; and the same would also have to hold for all states in a certain neighbourhood g of P_0; and for the states in this region that recur, the function would have to decrease again.

The same result can also be proved directly very simply. If the function S continually increased for all initial states in g, then it would also do so for all states in the larger region G, the future of g, and then according to (4) the n-fold integral over G,

$$\int S dx_1 dx_2 \ldots dx_n$$

must *continually increase*. But this is impossible since the region of integration, G, can change only by losing singular states that have no finite extension, so that the value of the integral must remain *constant*.

The interpretation and proof of this theorem become very clear for the case $n = 3$, when one takes the variables x_1, x_2, x_3 to be the coordinates of a mass-point in space. Then equations (1), in connection with (2) or (4), determine a *stationary flow of an incompressible fluid* in a *closed container*, if the quantities x_μ cannot increase without limit. A definite " state " corresponds here to a point in space, and a state varying in time corresponds to a moving point. The paths described by these fluid-points, the " stream-lines ", taken together form " stream-tubes " or " stream-filaments " according to whether they proceed from closed curves or from surface sections, and in the case of stationary motion they remain unchanged. Now it is intuitively obvious that in this case all the stream-filaments must run back on themselves, since the fluid streaming through the filament can neither break through the sides of the tube nor accumulate somewhere inside it. From this it follows that every finite fluid particle must always return again as closely as one likes to any position, if one makes the fluid filaments thin enough and waits a sufficiently long time. Of course there are also non-recurring singular stream-lines, for example those that *asymptotically approach* an immersed solid body or cavity between the other streamlines passing on either side and avoiding each other; but these cannot form a stream-filament of finite thickness. On the other hand, if the flow has a *velocity-potential*, then in a completely enclosed container this must necessarily be many-valued, while the function S mentioned above has to be single-valued. In the more general case $n > 3$, the same analogy is often of heuristic value; one can use the same terminology and interpret equations (1) and (2) or (4) as a " stationary flow of an incompressible fluid in a space of n dimensions ".

Our conclusion is therefore the following:

In a system of arbitrarily many mass-points, whose accelerations depend only on their position in space, there can be no " irreversible " processes for all initial states that occupy any region of finite extension, no matter how small, provided that the coordinates and velocities of the points can never exceed finite limits.

The theorem is even more general than this, for it holds in particular for an *arbitrary* mechanical system with generalized coordinates q_μ and corresponding momenta p_μ, whose equations of motion can be written in Hamiltonian form:

$$\frac{dp_\mu}{dt} = \frac{\partial H}{\partial q_\mu}, \frac{dq_\mu}{dt} = -\frac{\partial H}{\partial p_\mu}$$

which we may call a " conservative " system since all forces are

H

derivable from a potential, and hence the mechanical energy is conserved. In such a system it is obvious that

$$\frac{\partial}{\partial p}\frac{dq_\mu}{dt}+\frac{\partial}{\partial q}\frac{dq_\mu}{dt}=0$$

and by analogy with the relation (2) we see that all the consequences following from that relation retain their validity.

According to the mechanical theory, in its usual atomistic version, all of nature can be represented as a system of this type: all natural processes are nothing but motions of atoms or molecules which are treated either as points without extension, or as aggregates of such points, and which interact only with " central forces " derivable from a potential, which are independent of the velocities. It is precisely this assumption that one tries to use in the " kinetic theory of gases ", in that one treats the molecules of an " ideal gas " as centres of repulsion, or as elastic spheres, or (with Boltzmann) as elastic bodies of some other form, but in any case as a " conservative " system in the sense mentioned above. However, one also restricts oneself to short-range repulsive forces between pairs of molecules.

With these assumptions it would be impossible, on the basis of the foregoing considerations, to have " irreversible " processes, unless the molecules could be dispersed to infinite distances, or could eventually attain infinitely large velocities. (We exclude the possibility of a uniform progressive motion of the centre of mass of the system.) The former possibility is excluded by the special assumption that the system is surrounded by a solid container; and the latter is excluded by the principle of conservation of energy. For in order to attain an infinitely great kinetic energy, an infinite amount of work must be done, which could only occur in the case of very close approach of two attracting force-centres. However, we must assume on physical grounds that there can be only repulsive forces at very small distances between atoms.

Suppose that we have a gas enclosed in a solid container with elastic sides that are impenetrable to heat. In general there will indeed be an infinite manifold of initial states of the molecules for which the gas will undergo *permanent* changes of state, such as viscosity, heat condition, or diffusion. However, there will also be a much larger number of possible initial states, which can be reached by arbitrarily small displacements from the former states, and these states, instead of undergoing irreversible changes, will come back periodically to their initial states as closely as one likes in the sense described above. The same must also be true when the actual *physical* state—e.g. the temperature and the entropy—is not defined by an instantaneous *state of motion*, but rather by a finite *sequence of motions*, which in any case is determined by the initial state of motion, and must recur along with it.

Hence, in order to establish the general validity of the second law of thermodynamics, one would have to assume that only those initial states that lead to irreversible processes are actually realized in nature, despite their smaller number, while the other states, which from a mathematical viewpoint are more probable, actually *do not occur*.

Though such an assumption would be irrefutable, it would hardly correspond to our requirement for causality; and in any case the spirit of the mechanical view of nature itself requires that we should always assume that all *imaginable* mechanical initial states are physically *possible*, at least within certain limits, and certainly we must allow those states that constitute an overwhelming majority and deviate by an arbitrarily small amount from the ones that actually occur. Remember that, strictly speaking, all our laws of nature refer not to definite precise quantities or processes, which can never be observed with complete accuracy, but rather only to certain ranges of values, approximations, and probabilities, whereas singularities exist only as abstract limiting cases. The assumption being discussed here would therefore be quite unique in physics, and I do not believe that anyone would be satisfied with it for very long.

That *not all conceivable* initial states can correspond to the second law follows already from the fact that by reversing the velocity directions of all the molecules at an arbitrary time, the entire temporal course of a process must be reversed. In fact, this criticism of the mechanical explanation of irreversible processes has already been made before, and in the winter of 1894–95 there was an extended discussion of this question in *Nature*, stimulated by a remark of Culverwell, though it seems to me that no satisfactory solution was found.† It was not even proved that the physical state of a gas (which is the only significant quantity) must be the same for equal and opposite velocities of all the molecules; unless this is true we cannot speak here of an actual reversal of the process, and the possibility would still remain that, at least for an extended *region* of initial states, a continual increase of entropy could occur. Both of these are paradoxes against the argument given above, and they can be disposed of by applying Poincaré's theorem.

It is now necessary to formulate either the Carnot–Clausius principle or the mechanical theory in an essentially different way, or else decide to give up the latter theory altogether. Minor changes would not serve the purpose, it seems to me. For example, if one made the interatomic or intermolecular forces velocity-dependent, so that our theorem would no longer be applicable, then it would also be necessary to introduce (in order not to violate the conservation of energy) additional forces which do no work, and whose *direction* therefore is determined by the velocities. But then the force would no longer be one that acts between points, according to the law of action and reaction, which is essential to the entire atomic theory.

Regardless of whether it may be possible, by suitable alterations of the assumptions (as for example in Hertz's *Prinzipien der Mechanik*‡) to escape these contradictions, it is in any case *impossible* on the basis of the *present* theory to carry out a mechanical derivation of the second law without specializing the

† [E. P. Culverwell, *Nature* **50**, 617, **51**, 105, 246 (1894), **51**, 581 (1895); S. H. Burbury, *Nature* **51**, 78, 175, 320 (1894], **52**, 104 (1895); J. Larmor, *Nature* **51**, 152 (1894); L. Boltzmann, *Nature* **51**, 413, 581 (1895).]

‡ The Helmholtz theory of " cyclic systems " in its original form, on the other hand, would be affected by the consequences of Poincaré's theorem, since in the last analysis it also, if in another way, reduces to Hamilton's equations.

initial state. It is likewise impossible to prove that the well-known velocity distribution will be reached as a stationary final state, as its discoverers Maxwell and Boltzmann wished to do. I have not given a detailed examination of the various attempts at such a proof by Boltzmann and Lorentz,† since because of the difficulties of the subject I would rather explain as clearly as possible what can be proved rigorously and what seems to be of greatest importance, and thereby contribute to a renewed discussion and final solution of these problems.

---

† Recently collected in Boltzmann's *Vorlesungen über Gastheorie* I, 1896. [English translation, *Lectures on Gas Theory*, University of California Press, Berkeley, 1964; see also H. A. Lorentz, *Wien. Ber.* **95**, 115 (1887).]

8

Reply to Zermelo's Remarks on the Theory of Heat *

LUDWIG BOLTZMANN

SUMMARY

Poincaré's theorem, on which Zermelo's remarks are based [Selection 7], is clearly correct, but Zermelo's application of it to the theory of heat is not. The nature of the H-curve (entropy $vs.$ time) which can be deduced from the kinetic theory is such that if an initial state deviates considerably from the Maxwell distribution, it will tend toward that distribution with enormously large probability, and during an enormously long time will deviate from it by only vanishingly small amounts. Of course if one waits long enough, the initial state will eventually recur, but the recurrence time is so long that there is no possibility of ever observing it.

In contradiction to Zermelo's statement, the singular initial states which do not approach the Maxwell distribution are very small in number compared to those that do. Consequently there is no difficulty in explaining irreversible processes by means of the kinetic theory.

According to the molecular-kinetic view, the second law of thermodynamics is merely a theorem of probability theory. The fact that we never observe exceptions does not prove that the statistical viewpoint is wrong, because the theory predicts that the probability of an exception is practically zero when the number of molecules is large.

Clausius, Maxwell and others have already repeatedly mentioned that the theorems of gas theory have the character of

---

* [Originally published under the title: " Entgegnung auf die wärmetheoretischen Betrachtungen des Hrn. E. Zermelo ", *Annalen der Physik* **57**, pp. 773–84 (1896).]

statistical truths. I have often emphasized as clearly as possible† that Maxwell's law of the distribution of velocities among gas molecules is by no means a theorem of ordinary mechanics which can be proved from the equations of motion alone; on the contrary, it can only be proved that it has very high probability, and that for a large number of molecules all other states have by comparison such a small probability that for practical purposes they can be ignored. At the same time I have also emphasized that the second law of thermodynamics is from the molecular viewpoint merely a statistical law. Zermelo's paper‡ shows that my writings have been misunderstood; nevertheless it pleases me for it seems to be the first indication that these writings have been paid any attention in Germany.

Poincaré's theorem, which Zermelo explains at the beginning of his paper, is clearly correct, but his application of it to the theory of heat is not.

I have based the proof of Maxwell's velocity distribution law on the theorem that according to the laws of probability a certain quantity H (which is some kind of measure of the deviation of the prevailing state from Maxwell's) can only decrease for a stationary gas in a stationary container. The nature of this decrease will become most clear when one draws a graph (as I have done§) with time as abscissa and the corresponding values of H as ordinates, thus giving the so-called H-curve. (One may subtract off the minimum value H_{min} from all values of H.)

If one first sets the number of molecules equal to infinity and allows the time of the motion to become very large, then in the overwhelming majority of cases one obtains a curve which asymptotically approaches the abscissa axis.‖ The Poincaré theorem is not applicable in this case, as can easily be seen.

† L. Boltzmann, *Wien. Ber.* **75**, 67 (1877), **76**, 373 (1877), **78**, 740 (1878). " Der zweite Hauptsatz der Wärmetheorie ", lecture delivered 29 May 1866; *Almanach d. Wien. Akad.* [**36**, 225 (1886)], *Nature* **51**, 413 (1895). *Vorlesungen über Gastheorie* **1**, 42 (1896) [p. 58 in the English translation].

‡ Zermelo, *Ann. Physik* **57**, 485 (1896).

§ L. Boltzmann, *Nature* **51**, 413 (1895). [See also Figure on p. 244.]

‖ *Vorlesungen über Gastheorie* **1**, §5.

However, if one takes the time of the motion to be infinite, while the number of molecules is very large but not actually infinite, then the H-curve has a different appearance. As I have already shown (footnote §, page 219), it almost always runs very close to the abscissa axis. Only very rarely does it rise up above this axis; we call this a peak, and indeed the probability of a peak decreases very rapidly as the height of the peak increases. At those times when the ordinate of the H-curve is very small, Maxwell's distribution holds almost exactly; but significant deviations occur at high peaks of the H-curve. Zermelo thinks that he can conclude from Poincaré's theorem that it is only for certain singular initial states, whose number is infinitesimal compared to all possible initial states, that the Maxwell distribution will be approached, while for most initial states this law is not obeyed. This seems to me to be incorrect. It is just for certain singular initial states that the Maxwell distribution is never reached, for example when all the molecules are initially moving in a line perpendicular to two sides of the container. For the overwhelming majority of initial conditions, on the other hand, the H-curve has the character mentioned above.

If the initial state lies on an enormously high peak, i.e. if it is completely different from the Maxwellian state, then the state will approach this velocity distribution with enormously large probability, and during an enormously long time it will deviate from it by only vanishingly small amounts. Of course if one waits an even longer time, he may observe an even higher peak, and indeed the initial state will eventually recur; in a mathematical sense one must have an infinite time duration infinitely often.

Zermelo is therefore completely correct when he asserts that the motion is periodic in a mathematical sense; but, far from contradicting my theorem, this periodicity is in complete harmony with it. One should not forget that the Maxwell distribution is not a state in which each molecule has a definite position and velocity, and which is thereby attained when the position and velocity of each molecule approach these definite values asymptotically. For a finite number of molecules the Maxwell distribution can never be true exactly, but only to a high degree of approximation. It is in no way a special singular distribution which is to be contrasted to infinitely many more non-Maxwellian distributions; rather it is characterized by the fact that by far the largest number of possible velocity distributions have the characteristic properties of the Maxwell distribution, and compared to these there are only a relatively small number of possible distributions that deviate significantly from Maxwell's. Whereas Zermelo says that the number of states that finally lead to the Maxwellian state is small compared to all possible states, I assert on the contrary that by far the largest number of possible states are " Maxwellian " and that the number that deviate from the Maxwellian state is vanishingly small.†

For the first molecule, any position in space, and any values of its velocity comqonents consistent with conservation of total energy, are equally probable.

If one combines all states of all molecules, then he obtains in almost every case the Maxwell distribution, to a high degree of approximation.   Only a few combinations give a completely different distribution of states.

An analogy for this is provided by the theory of the method of least squares, where one assumes that each elementary error is equally likely to have a positive or equal negative value; it is then proved that if one combines all possible values of the elementary errors in all possible ways, the great majority of combinations will obey the Gaussian law of errors, and for relatively few combinations will there be significant deviations; the deviations are not impossible, but they are very unlikely.

An even simpler example is provided by the game of dice. In 6000 throws with the same dice one might obtain 1000 one's, 1000 two's, and so forth, not because any such random sequence of throws is more probable than a series of 6000 one's, but rather because there are many more possible combinations corresponding

---

† For the definition of equally probable states, see my papers cited on previous pages.

to an equal number of one's, two's, etc., than corresponding to all one's.

The theory of probability therefore leads to the result (as is well known) that a recurrence of an initial state is not mathematically impossible, and indeed is to be expected if the time of the motion is sufficiently long, since the probability of finding a state very close to the initial state is very small but not zero. The consequence of Poincaré's theorem—that, apart from a few singular initial states, a state very close to the initial state must eventually occur after a very long time—is therefore in complete agreement with my theory.

It is only the conclusion that the mechanical viewpoint must somehow be changed or even given up that is incorrect. This conclusion would be justified only if the mechanical viewpoint led to some consequence that was in contradiction to experience. This would only be the case, however, if Zermelo could prove that the duration of the period of time after which the previous state of the gas must recur according to Poincaré's theorem has an observable length. It should indeed be obvious that if a trillion tiny spheres, each with a high velocity, are initially collected together in one corner of a container with absolutely elastic walls, then in a very short time they will be uniformly distributed throughout the container; and that the time required for all their collisions to have compensated each other in such a way that they all come back to the same corner, must be so large that no one will be present to observe it. Though it seems unnecessary, I have estimated the magnitude of this time in the appendix, and the value obtained is comfortingly large. Though this calculation makes no pretense to accuracy, it still shows that it cannot be proved from Poincaré's theorem that the theoretical existence of a recurrence time involves any contradiction with experience, since the length of this time makes any attempt to observe it ridiculous. The states that we observe all fall in the intermediate time between the beginning and end of the cycle, so that Poincaré's theorem does not exclude states that approximate with arbitrary accuracy the Maxwellian state.

Zermelo's case is therefore only one of many cases (and indeed one that does exceptionally little harm to gas theory) where a state that is theoretically only very improbable must be considered as never occuring in practice. Thus for example in oxyhydrogen gas at ordinary temperatures there must be occasional collisions of two or three molecules with very high velocities; if these were not excluded, oxyhydrogen gas would turn into water at ordinary temperatures.

To give another example, the case that during one second no molecule of a gas collides with a piston is only very improbable but not impossible.

The time that one must wait for a measurable amount of water to be produced from oxyhydrogen gas at ordinary temperature, or for the pressure on a piston to decrease by a measurable amount from its average value, is not as long as a recurrence time, but it is still sufficiently long to preclude observation.   An argument against the kinetic theory can be derived from such considerations only when such phenomena fail to appear in a period of time for which calculation indicates that they should appear. This does not seem to be the case; on the contrary, for temperatures lower than the conversion temperature, actual traces of chemical conversion can be found; likewise, it is observed that very small particles in a gas execute motions which result from the fact that the pressure on the surface of the particles may fluctuate.

Thus when Zermelo concludes, from the theoretical fact that the initial states in a gas must recur—without having calculated how long a time this will take—that the hypotheses of gas theory must be rejected or else fundamentally changed, he is just like a dice player who has calculated that the probability of a sequence of 1000 one's is not zero, and then concludes that his dice must be loaded since he has not yet observed such a sequence!

The foregoing remarks are intimately connected with my interpretation of the second law of thermodynamics in the papers cited above. According to the molecular-kinetic view, this law is merely a theorem of probability theory. According to this view, it cannot be proved from the equations of motion that all phenomena must evolve in a certain direction in time. For all phenomena where only visible motion occurs, so that the body always moves as a whole, both directions must be equivalent. On the other hand, when the motion involves a very large number of very small molecules, then there must be (aside from a small number of exceptional cases) a progression from less probable to more probable states, and therefore a continual change in a definite direction, such as, in a gas, the evolution toward a Maxwellian distribution. On the other hand, when it is a question of the motions of individual molecules, this would no longer be expected.

The first and second cases are confirmed by experience; the third case has not yet been realized. Its possibility is hence neither proved nor disproved. Famous scientists, such as Helmholtz,† have believed this, and as I have tried to indicate in my book on gas theory,‡ the opinion that the second law is merely a statistical law is not only not contradicted by the facts but agrees rather well with them. Gibbs§ also arrived, by considering purely empirical facts, at the following conclusion: " The impossibility of an incompensated decrease of entropy seems to be reduced to an improbability ".

We therefore arrive at the following result: if one considers heat to be molecular motion which takes place according to the general equations of mechanics, and assumes that the complexes of bodies that we observe are at present in very improbable states, then he can obtain a theorem which agrees with the second law for phenomena observed up to now.

Of course as soon as one observes bodies of such small size that they contain only a few molecules, the theorem will no longer be valid. However, since no experiments have yet been done on such small bodies, the assumption does not contradict our present

---

† Helmholtz, *Berlin Ber.* **17**, 172 (1884).

‡ L. Boltzmann, *Vorlesungen über Gastheorie* 1, p. 61 [p. 75 in the English translation].

§ Gibbs, *Trans. Conn. Acad.* **3**, 229 (1875); p. 198 in Ostwald's German edition.

experience; indeed, the experiments that have been done on small particles in gases are favourable to the assumption, although we can hardly say that we have an experimental proof of it yet.

When the bodies in question contain many molecules, there must occur very small deviations from this theorem, since the number of molecules is not infinite. But these deviations could only add up to an observable value in a very long period of time, so that this consequence of atomistics cannot be tested by experiment. This is all the more true since gas theory claims to give only an approximate description of reality. Perturbations experienced by the molecules as a result of the aether or the electrical properties of the molecules, etc., must be left out of the theory because of our complete ignorance concerning such effects. There is no such thing as an absolutely smooth wall; on the contrary, every gas is really interacting with the entire universe, and hence the validity of the kinetic theory is not destroyed by small deviations from experience.

An answer to the question—how does it happen that at present the bodies surrounding us are in a very improbable state—cannot be given, any more than one can expect science to tell us why phenomena occur at all and take place according to certain laws.

Gas theory is not to be confused with the theory of central forces—i.e. with the hypothesis that all natural phenomena can be explained by means of central forces between mass points—since gas theory does not assume that either the properties of the aether or the internal constitution of molecules can be explained by centres of force, but only that for the interaction of two molecules during a collision the Lagrange equations of motion are valid with sufficient accuracy for the explanation of thermal phenomena.

A consequence of the Poincaré theorem may still be used against the theory of central forces with respect to the properties of the entire universe. One may say that according to Poincaré's theorem the entire universe must return to its initial state after a sufficiently long time, and hence there must be times when all processes take place in the opposite direction. How shall we decide, when we leave the domain of the observable, whether the age of the universe, or the number of centres of force which it contains is infinite? Moreover, in this case the assumption that the space available for the motion, and the total energy, are finite, is questionable. The assumption of the unlimited validity of the irreversibility principle, when applied to the universe for an infinitely long period of time, leads (as is well known) to the scarcely more attractive consequence that, when all irreversible processes have been played out, the universe will continue to exist without any events, or all events will gradually disappear. Just as it would be wrong to deduce from this the incorrectness of the irreversibility principle, so it would also be wrong to suppose that it proves anything against atomistics.

All the paradoxes raised against the mechanical viewpoint are therefore meaningless and based on errors. However, if the difficulties offered by the clear comprehension of gas-theoretic theorems cannot be overcome, then we should in fact follow the suggestion of Zermelo and decide to give up the theory entirely.

Appendix

We assume a container of volume 1 cc. In this container there will be about a trillion ($= n$) molecules of air at ordinary density. The velocity of each molecule will initially be 500 metres per second. The average distance between the centres of two neighbouring molecules is about 10^{-6} cm.

We now construct around the midpoint of each molecule a cube of edge-length 10^{-7} cm, which we call the initial space of the molecule in question. We also construct a velocity diagram by representing the velocity of each molecule by a line from the origin with the appropriate magnitude and direction. The end-point of this line is called the velocity point of the molecule. Here we divide the entire infinite space into cubes of 1 metre edge length, which we call the elementary cubes. The elementary cube in which the velocity point of a molecule is found initially will be called the initial space of its velocity point.

We now ask after how long a time, according to Poincaré's theorem, will the centres and velocity points of all the molecules return simultaneously to their initial spaces? Note that we do not require exact recurrence, since we accept the velocity state of a molecule as being the same as its initial state if its velocity components return to values that differ by no more than 1 metre from their original values.

We assume that each molecule experiences 4.10^9 collisions per second. It then follows that there will be in all about $b = 2.10^{27}$ collisions per second in the gas. In such a collision, the velocity points of two molecules will generally be displaced to different elementary cubes. According to Poincaré's theorem the original state does not have to recur until the velocity points have gone through all possible combinations of the elementary cubes.

The first molecule can have all possible velocities from zero up to $(500.10^9 = a)$ m/sec. If it has velocity v_1 m/sec, then the second can have all possible velocities from zero up to $\sqrt{a^2 - v_1^2}$ m/sec, and so forth.

The number of possible combinations of all the velocity points in the different elementary cubes is therefore:

$$N = (4\pi)^{n-1} \int_0^a v_1^2 dv_1 \int_0^{\sqrt{a^2-v_1^2}} v_2^2 dv_2 \ldots \int_0^{\sqrt{a^2-v_1^2\ldots v_{n-2}^2}} v_{n-1}^2 dv_{n-1}$$

$$= \frac{\pi^{(3n-3)/2} a^{3(n-1)}}{2.3.4. \ldots [3(n-1)/2]}$$

or

$$\frac{2.(2\pi)^{(3n-4)/2} a^{3(n-1)}}{3.5.7 \ldots 3(n-1)}$$

according as n is odd or even.

Since each of these combinations lasts on the average $1/b$ seconds, all of them will be gone through in N/b seconds. After this time all molecules except one must have come back to their original velocity state. The velocity direction of this last molecule is not restricted, nor is the position of the centre of any of the molecules. In order to make the state the same as the original one, the midpoint of each molecule must also return to its initial space, so that the above number must again be multipled by another number of similar magnitude.

Though the number N/b is enormous, one can obtain some idea of its magnitude by noting that it has many trillions of digits. For comparison, suppose that every star visible with the best telescope has as many planets as does the sun, and on each planet live as many men as are on the earth, and each of these men lives a trillion years; then the total number of seconds that they all live will still have less than 50 digits.

If the gas molecules were initially distributed uniformly throughout the container, and all of them had the same velocity, then after only a hundred-millionth of a second they would already have nearly a Maxwellian velocity distribution. Comparison of these numbers shows, on the one hand, how small a fraction of the total number of possible state distributions is made up of those that deviate noticeably from the Maxwell distribution; and on the other hand, how certain are such theorems that theoretically are merely probability laws but in practice have the same significance as laws of nature.

Vienna, March 20, 1896.

9

On the Mechanical Explanation of
Irreversible Processes *

ERNST ZERMELO

SUMMARY

Boltzmann has conceded [Selection 8] that the commonly accepted version of the second law of thermodynamics is incompatible with the mechanical viewpoint. Whereas the author holds that the former, a principle that summarizes an abundance of established experimental facts, is more reliable than a mathematical theorem based on unverifiable hypotheses, Boltzmann wishes to preserve the mechanical viewpoint by changing the second law into a " mere probability theorem ", which need not always be valid.

Boltzmann's assertion, that the statistical formulation of the second law is really equivalent to the usual one, is based on postulated properties of the H-curve which he has not proved, and which seem to be impossible. His argument that any arbitrarily chosen initial state will probably be a maximum on the H-curve, if it were valid, would prove that the H-curve consists entirely of maxima, which is nonsense.

The only way that the mechanical theory can lead to irreversibility is by the introduction of a new physical assumption, to the effect that the initial state always corresponds to a point at or just past the maximum on the H-curve; but this would be assuming what was supposed to be proved.

My paper in the March issue of this Journal, " On a theorem of dynamics and the mechanical theory of heat,"† has drawn from Herr Boltzmann an immediate reply,‡ in which I find a confirma-

---

* [Originally published under the title: " Ueber mechanische Erklärungen irreversibler Vorgänge ", *Annalen der Physik* **59**, 793–801 (1896).]

† E. Zermelo, *Ann. Physics* **57**, 485 (1896). [Selection 7]

‡ L. Boltzmann, *Ann. Physik* **57**, 773 (1896). [Selection 8]

J*

tion of my own views rather than a contradiction. Not only does Herr Boltzmann recognize that the basic theorem of Poincaré is "obviously correct", but he also concedes that it is applicable to a closed system of gas molecules in the sense of the kinetic theory. Indeed, in such a system all processes are *periodic* from a mathematical viewpoint, hence *not irreversible* in the strict sense, so that one may not assert that there is an actual progressive increase of entropy as the second law, in its usual meaning, would require. To prove this, and thereby to obtain a firm basis for the discussion of the principal questions, was the purpose of my paper; at the time I was not familiar with Herr Boltzmann's investigations of gas theory, but I still think that this general clarification was not at all superfluous.

The " necessity of making a fundamental modification either in the Carnot–Clausius principle or the mechanical viewpoint " which I asserted is therefore conceded, and it remains a matter of personal opinion which of these possibilities is to be chosen. As for me (and I am not alone in this opinion), I believe that a single principle summarizing an abundance of established experimental facts is more reliable than a mathematical theorem, which by its nature represents only a theory which can never be directly verified; I prefer to give up the theorem rather than the principle, if the two are inconsistent.

Herr Boltzmann, however, will not modify the ordinary mechanical viewpoint, and instead wishes to change the second law into a " mere probability theorem " which is not valid at all times. Yet he asserts that this change, whose *principal* meaning he does not misunderstand, is really unimportant, and that " in practice " his two formulations are " completely equivalent ". Let us see how far he has succeeded in proving this.

It is undoubtedly correct, as Boltzmann emphasizes, that for a very large number of molecules in a finite volume the average duration of the Poincaré period, the time after which a state will recur, is much too large for us to expect to make a direct *observation* of the theoretical periodicity. However, his numerical estimate, which is based on a single exceptional initial state with a completely determined molecular configuration, is not conclusive. In practice one is interested in a " physical state " which can be realized by many possible combinations, and can therefore recur very much earlier. Moreover, for my purposes it is sufficient to prove the recurrence of any other state with the same or a smaller value of the entropy; the periods of recurrence of such individual values of the entropy S will of course vary, but on the whole they no longer come out to be so " comfortingly " large. Nevertheless there are functions whose periodicity is beyond observation, and the entropy function might be one of them.

For such a function it can of course happen that it *appears* to be continually increasing, since the decreasing branch of the curve, which is theoretically always present, begins so much later that it does not need to be considered. Yet it by no means follows from this that there are functions for which one *always* observes the increasing and not the decreasing part, which is the property that the mechanical analog of the entropy function must have. It is not satisfactory simply to accept this property as a fact for a particular type of initial state that we can observe at present, for it is not a question of a variable which is just observed once (as for example the eccentricity of the earth's orbit) but of the entropy of *any* *arbitrary* system free of external influences. How does it happen, then, that in such a system there always occurs only an *increase* of entropy and *equalization* of temperature and concentration differences, but never the reverse? And what right do we have to expect this behaviour to continue, at least for the immediate future? A satisfactory answer to this question must be given, if we are to accept a mechanical analog of the second law.

It seems to me that probability theory cannot help here, since every increase corresponds to a later decrease, and both must be equally probable or at least have probabilities of the same order of magnitude. My opinion, in agreement with Poincaré's definition,† is that the probability of occurrence of a certain property of the molecular states, for example for a definite value of the function S,

---

† H. Poincaré, *Acta Math.* **13**, 71 (1890). [Selection 5, p. 199]

can be measured only by the "extension"† γ of the "region" g of all possible states which have this property, divided of course by the total extension Γ of the region G containing all possible states. But since according to Liouville's theorem each extension γ is independent of time, any such value of a function must have the same probability at a later time as at the initial time, and no overall increase or decrease is to be expected on the grounds of probability theory.‡

Herr Boltzmann proceeds in a different way. He assumes a function H whose curve, drawn with the time t as abscissa, runs in general very close to the t-axis but occasionally has elevations or "peaks". The larger the peaks are, the more improbable they are, and the less often they occur.§ I cannot find that he has actually *proved* this property from his other definition of the H-function. According to my definition, probability and duration of a state are not identical. Nevertheless, functions of the indicated nature may exist. He further assumes that the H-function has initially an unusually large value H_0, corresponding to a peak, but soon passes this peak and decreases almost to zero. Finally, it runs very close to the abscissa axis for a very long time. This limiting value zero of the H-function corresponds to a velocity distribution expressed by Maxwell's law, so that the properties of this H-curve provide an explanation of the probability-theoretic meaning of the distribution law, which however I do not dispute. The law does not represent a "stationary *final* state" in the strict sense, since the curve eventually rises to new peaks after a long time. Herr Boltzmann himself considers the Maxwellian state to be the "*final* state" only in an empirical or approximate sense, and it seems to me that *this* assertion does not follow sufficiently clearly from his earlier writings.

---

† E. Zermelo, *Ann. Physik* **57**, 485 (1896). [Selection 7, p. 210]

‡ [This argument is developed in more detail by Gibbs in his discussion of the generalized H-theorem: see *Elementary Principles in Statistical Mechanics* (Scribner, New York, 1902) Chapter XII.]

§ L. Boltzmann, *Ann. Physik* **57**, 773 (1896). [Selection 8, p. 220]

But it is not here a question of Maxwell's law, but of whether an analogy exists between the properties of the H-curve and the second law of thermodynamics; it is this analogy that I dispute. It is not sufficient to show that all perturbations *finally* relax to a long-lasting equilibrium state; rather it is necessary to show that changes always take place in the same sense, in the direction of equalization; that the H-function *always only* decreases during observable times, or at least that there can only be very small, practically unnoticeable increases, which will always be immediately washed out by stronger decreases. In my opinion this proof is as little possible for the H-function as for any other function. Clearly the initial state, whose probability can depend only on the initial value H_0, can just as well lie on a rising as a falling branch of the curve, and in the former case there must first be an *increase*, which can last just as long as the subsequent decrease. For this period we have $H > H_0$. Each observed decrease $H_1 \ldots H_2$ in the falling branch corresponds to an equally great increase $H_2 \ldots H_1$ in the rising branch, and the process is no more likely to begin in one way than the other. If the increase takes place in a shorter time and is hence less probable than the decrease—an assumption for which there is no basis in the theory—then it would still have to be *steeper* and therefore should be given just as much weight.

Herr Boltzmann's assertion, if I have understood it correctly,† is that the initial state has a fairly large H-value, say $H_0 > H'$, on a peak which is not too large (so that it does not have too small a probability) and as a rule must represent a *maximum*, so that of course one always observes only the decreasing branch. I cannot conceive of such a curve. Suppose for the sake of argument that the intersections of the H-curve with a line parallel to the time-axis at a height $H = H_0$ are mostly maxima, and that $H_0 > H'$. But where are the other points on the peak ($H > H'$) which are *not* maxima? Are they in fact in the minority compared to the

---

† L. Boltzmann, *Vorlesungen über Gastheorie* **1,** 44 (1896) [p. 59 in the English translation].

maxima? It is clear that this argument can make sense only if the maxima are considered not as mathematical points but as having a certain breadth, i.e. a certain time-duration. But then for any initial state the value of the function will remain constant for a longer or shorter time, thereby representing a sort of labile equilibrium; whereas according to experience, for example in the case of heat conduction, the process of equalization begins more rapidly, the greater the initial temperature differences are, that is, the further the initial state is from the stable equilibrium state.

Aside from this, I do not understand what the *initial* state has to do with the argument, except for its property of having a small probability, which it shares with the neighbouring states. Herr Boltzmann assumes that the entire H-curve, and therefore the collection of all states through which the system passes, is *given* and now asks for the probability of a certain initial state, i.e. the place on the curve where the system actually begins to move, without any external forces being present. But, as experience teaches, there is no procedure available for producing *any arbitrary* initial state by an appropriate action and then isolating the system and letting it run by itself; one cannot make any arbitrary state P_0 the initial state. If this were true, then the system would actually pass through all the states P that follow P_0 in the series, while the previous states could only be added mathematically. Now if the above argument were correct, and the initial state represents a maximum of the H-function in most cases, then the same must also be true of *all other* states for which H exceeds H', since any other state could be chosen as initial state. Moreover, the whole probability argument is just as applicable to any arbitrary state as to the initial state. All these states must therefore represent maxima, and the curve must consist purely of maxima above a certain height. This is nonsense, since the function cannot be constant. Therefore in order to obtain an approximate empirical analog of the entropy theorem, it is not sufficient to assume that the initial state is extraordinarily improbable; rather one must add the *new assumption*, that at the beginning the H-curve has a maximum or has just passed a maximum. But as long as one cannot make comprehensible the *physical origin* of the initial state, one must merely assume what one wants to prove; instead of an explanation one has a renunciation of any explanation.

I have therefore not been able to convince myself that Herr Boltzmann's probability arguments, on which " the clear comprehension of the gas-theoretic theorem "† is supposed to rest, are in fact able to dispel the doubts of a mechanical explanation of irreversible processes based on Poincaré's theorem, even if one renounces the strict irreversibility in favour of a merely empirical one. Indeed it is clear *a priori* that the probability concept has nothing to do with time and therefore cannot be used to deduce any conclusions about the *direction* of irreversible processes. On the contrary, any such deduction would be equally valid if one interchanged the initial and final states and considered the *reversed* process running in the opposite direction. Hence, the following dice game is more relevant than the example introduced by Herr Boltzmann. Two dice-players, let us suppose, have made the observation that dice they obtain from a certain source always behave in a certain way when they first start to play with them. One particular face, say the one, always comes up first. In the first 600 throws, the one comes up 200 times rather than 100 times. However, in the next 6000 throws the ones are less frequent, and after the game has continued a long time they find that one comes up on the average only 100 times out of 6000, like all the other numbers. The first player sees nothing strange in this behaviour, since the laws of probability theory are supposed to apply to very long games. But the second player says: No! This dice must be false, and it is only through long use that it gradually regains its proper condition—the latter interpretation corresponds to my own opinion.

Not only is it impossible to explain the general *principle* of irreversibility, it is also impossible to explain individual irreversible

---

† L. Boltzmann, *Ann. Physik* **57**, 773 (1896). [Selection 9, p. 226]

processes themselves without introducing new physical assumptions, at least as far as the time-direction is concerned. In particular, the differential equation for heat conduction and diffusion is

$$\frac{\partial u}{\partial t} = a^2 \frac{\partial^2 u}{\partial x^2}$$

and this equation can only represent irreversible processes. The attempt to derive this equation purely from the basic equations of mechanics, together with probability assumptions, which has been been made for example by Clausius, Maxwell, and Boltzmann, cannot reach its goal, since it is an impossible undertaking, and an apparent success can only rest on an error of deduction. The major fallacy in the methods heretofore applied seems to me to be the unprovable (because untrue) assumption that the molecular state of a gas is always, in Boltzmann's expression, " disordered " and that all possible directions and combinations are equivalent, if one can say nothing definite about the true state, which must nevertheless depend on the " ordered " initial state.† Probability theory justifies such assumptions to a certain extent for the *initial state*, at most; the probability of a later state, however, and therefore the process itself, must always first be expressed in terms of the corresponding initial state, and only then can one decide on the permissibility of such averaging assumptions. The difficulty of carrying out investigations *rigorously* from the viewpoint of probability theory may be very great, but they do not seem to me to be insurmountable. In any case such investigations cannot by themselves correct the errors of the " statistical method " used up to now; questions of principle, such as those under discussion here, require arguments whose mathematical validity is beyond question. For the present I must restrict myself to these remarks; I hope later to return to a more explicit treatment of these methodological questions.

The great successes of the kinetic theory of gases in the explanation of *equilibrium* properties do not entail its applicability to

---

† Boltzmann, *Gastheorie* **1**, 21 (1896) [p. 40 in the English translation].

time-dependent processes also, for the two are separate subjects; while in the former case the theory frequently gives us a correct and valuable picture, in the latter case, especially where it is a question of the explanation of irreversible processes, it must necessarily fail unless completely new assumptions are added to it.

Berlin, 15 September 1896.

10

On Zermelo's Paper "On the Mechanical Explanation of Irreversible Processes" *,†

LUDWIG BOLTZMANN

SUMMARY

The second law of thermodynamics can be proved from the mechanical theory if one assumes that the present state of the universe, or at least that part which surrounds us, started to evolve from an improbable state and is still in a relatively improbable state. This is a reasonable assumption to make, since it enables us to explain the facts of experience, and one should not expect to be able to deduce it from anything more fundamental.

The applicability of probability theory to physical situations, which is disputed by Zermelo, cannot by rigorously proved, but the fact that one never observes those events that theoretically should be quite rare is certainly not a valid argument against the theory.

One may speculate that the universe as a whole is in thermal equilibrium and therefore dead, but there will be local deviations from equilibrium which may last for the relatively short time of a few eons. For the universe as a whole, there is no distinction between the " backwards " and " forwards " directions of time, but for the worlds on which living beings exist, and which are therefore in relatively improbable states, the direction of time will be determined by the direction of increasing entropy, proceeding from less to more probable states.

I will be as brief as possible without loss of clarity.

§1. The second law will be explained mechanically by means of of assumption A (which is of course unprovable) that the universe,

---

* [Originally published under the title: " Zu Hrn. Zermelo's Abhandlung Über die mechanische Erklärung irreversibler Vorgange ", *Annalen der Physik* **60**, pp. 392–8 (1897).]

† E. Zermelo, *Ann. Physik* **59**, 793 (1896). [Selection 9]

considered as a mechanical system—or at least a very large part
of it which surrounds us—started from a very improbable state,
and is still in an improbable state. Hence, if one takes a smaller
system of bodies in the state in which he actually finds them, and
suddenly isolates this system from the rest of the world, then the
system will initially be in an improbable state, and as long as the
system remains isolated it will always proceed toward more
probable states. On the other hand, there is a very small probability that the enclosed system is initially in thermal equilibrium,
and that while it remains enclosed it moves far enough away from
equilibrium that its entropy decrease is noticeable.

The question is not what will be the behaviour of a completely
arbitrary system, but rather what will happen to a system existing
in the present state of the world. The initial state precedes the
later states, so that Zermelo's conclusion that all points of the
H-curve must be maxima is invalid. Hence, it turns out that
entropy always increases, temperature and concentration differences are always equalized, that the initial value of H is such
that during the time of observation it almost always decreases, and
that initial and final states are not interchangeable, in contradiction to Zermelo's assertions. Assumption A is a comprehensible
physical explanation of the peculiarity of the initial state, consistent with the laws of mechanics; or better, it is a unified viewpoint corresponding to these laws, which allows one to predict
the type of peculiarity of the initial state in any special case; for
one can never expect that the explanatory principle must itself be
explained.

On the other hand, if we do not make any assumption about the
present state of the universe, then of course we cannot expect to
find that a system isolated from the universe, whose initial state is
completely arbitrary, will be in an improbable state initially rather
than later. On the contrary it is to be expected that at the moment
of separation the system will be in thermal equilibrium. In the few
cases where this does not happen, it will almost always be found
that if the state of the isolated system is followed either backwards
or forwards in time, it will almost immediately pass to a more

probable state. Much rarer will be the cases in which the state becomes still more improbable as time goes on; but such cases will be just as frequent as those where the state becomes more improbable as one follows it backwards in time.

§2. The applicability of probability theory to a particular case cannot of course be proved rigorously. If, out of 100,000 objects of a certain kind, about 100 are annually destroyed by fire, then we cannot be sure that this will happen next year. On the contrary, if the same conditions could be maintained for $10^{10^{10}}$ years, then during this time it would often happen that all 100,000 objects would burn up on the same day; and likewise there will be entire years during which not a single object is damaged. Despite this, every insurance company relies on probability theory.

It is even more valid, on account of the huge number of molecules in a cubic millimetre, to adopt the assumption (which cannot be proved mathematically for any particular case) that when two gases of different kinds or at different temperatures are brought in contact, each molecule will have all the possible different states corresponding to the laws of probability and determined by the average values at the place in question, during a long period of time. These probability arguments cannot replace a direct analysis of the motion of each molecule; yet if one starts with a variety of initial conditions, all corresponding to the same average values (and therefore equivalent from the viewpoint of observation), one is entitled to expect that the results of both methods will agree, aside from some individual exceptions which will be even rarer than in the above example of 100,000 objects all burning on the same day. The assumption that these rare cases are not observed in nature is not strictly provable (nor is the entire mechanical picture itself) but in view of what has been said it is so natural and obvious, and so much in agreement with all experience with probabilities, from the method of least squares to the dice game, that any doubt on this point certainly cannot put in question the validity of the theory when it is otherwise so useful.

It is completely incomprehensible to me how anyone can see a refutation of the applicability of probability theory in the fact that some other argument shows that exceptions must occur now and then over a period of eons of time; for probability theory itself teaches just the same thing.

§3. Let us imagine that a partition which separates two spaces filled with different kinds of gas is suddenly removed. One could hardly find another situation (at least one in which the method of least squares is applicable) where there are so many independent causes acting in such different ways, and in which the application of probability theory is so amply justified. The opinion that the laws of probability are not valid here, and that in most cases the molecules do not diffuse, but instead a large part of the container has significantly more nitrogen, and another part has significantly more oxygen, cannot be disproved, even if I were to calculate exactly the motions of trillions of molecules in millions of different special cases. Nevertheless this opinion certainly does not have enough justification to cast doubt on the usefulness of a theory that starts from the assumption of the applicability of probability theory and draws the logical consequence from this assumption.

Poincaré's theorem does not contradict the applicability of probability theory but rather supports it, since it shows that in eons of time there will occur a relatively short period during which the state probability and the entropy of the gas will significantly decrease, and that a more ordered state similar to the initial state will occur. During the enormously long period of time before this happens, any noticeable deviation of the entropy from its maximum value is of course very improbable; however, a momentary increase or decrease of entropy is equally probable.

It is also clear from this example that the process goes on irreversibly during observable times, since one intentionally starts from a very improbable state. In the case of natural processes this is explained by the assumption that one isolates the system of bodies from the universe which is at that time in a very improbable state as a whole.

This example of two initially unmixed gases gives us incidentally a possible way of imagining the initial state of the world. For if in the example we isolate the gas found in a smaller space soon after the beginning of the diffusion from the rest of the gas, we will have the asymmetry with respect to forward and backward steps in time as in the isolated system of bodies mentioned in §1.

§4. I myself have repeatedly warned against placing too much confidence in the extension of our thought pictures beyond the domain of experience, and I am aware that one must consider the form of mechanics, and especially the representation of the smallest particles of bodies as mass-points, to be only provisionally established. With all these reservations, it is still possible for those who wish to give in to their natural impulses to make up a special picture of the universe.

One has the choice of two kinds of pictures. One can assume that the entire universe finds itself at present in a very improbable state. However, one may suppose that the eons during which this improbable state lasts, and the distance from here to Sirius, are minute compared to the age and size of the universe. There must then be in the universe, which is in thermal equilibrium as a whole and therefore dead, here and there relatively small regions of the size of our galaxy (which we call worlds), which during the relatively short time of eons deviate significantly from thermal equilibrium. Among these worlds the state probability increases as often as it decreases. For the universe as a whole the two directions of time are indistinguishable, just as in space there is no up or down. However, just as at a certain place on the earth's surface we can call " down " the direction toward the centre of the earth, so a living being that finds itself in such a world at a certain period of time can define the time direction as going from less probable to more probable states (the former will be the " past " and the latter the " future ") and by virtue of this definition he will find that this small region, isolated from the rest of the universe, is " initially " always in an improbable state. This viewpoint seems to me to be the only way in which one can understand the validity of the second law and the heat death of each individual world without invoking an unidirectional change of the entire universe from a definite initial state to a final state. The objection that it is uneconomical and hence senseless to imagine such a large part of the universe as being dead in order to explain why a small part is living—this objection I consider invalid. I remember only too well a person who absolutely refused to believe that the sun could be 20 million miles from the earth, on the grounds that it is inconceivable that there could be so much space filled only with aether and so little with life.

§5. Whether one wishes to indulge in such speculations is of course a matter of taste. It is not a question of choosing as a matter of taste between the Carnot–Clausius principle and the mechanical theory. The importance of the former, as the simplest expression of the facts so far observed, is not in dispute. I assert only that the mechanical picture agrees with it in all actual observations. That it suggests the possibility of certain new observations—for example, of the motion of small particles in liquids and gases, and of viscosity and heat conduction in very rarefied gases, etc.—and that it does not agree with the Carnot–Clausius principle on some unobservable questions (for example the behaviour of the universe or a completely enclosed system during an infinite period of time), may be called a difference in principle, if you like. In any case it provides no basis for giving up the mechanical theory, as Herr Zermelo would like to do, if it cannot be changed in principle (which one should not expect). It is precisely this difference that seems to me to indicate that the universality of our thought-pictures will be improved by studying not only the consequences of the principle in the Carnot–Clausius version but also in the mechanical version.

Appendix

§6. I have always measured the probability of a state, independently of its temporal duration, by the " extension γ " (as Zermelo calls it) of its corresponding region, and I used the Liouville theorem in this connection 30 years ago.† The Maxwellian state

† See especially *Wien. Ber.* **58,** 517 (1868); **63,** 679, 712 (1871); **66,** (1872); **76,** 373 (1877). I have given there the proof of the above-mentioned theorem, which there is not space to repeat here.

is simply the most probable because it can be realized in the largest
number of ways. The total extension γ of the region of all those
states for which the velocity distribution is approximately given by
the Maxwell distribution is therefore much greater than the total
extension of the regions of all other states. It was only to illustrate
the relation between the temporal course of the states and their
probabilities that I represented the reciprocal value of this

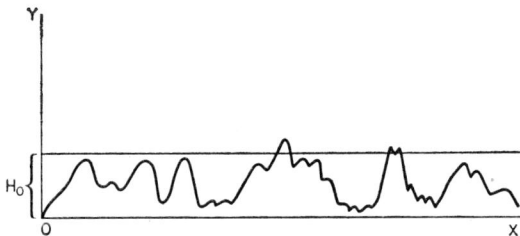

Fig. 1.

probability for the different successive states by the H-curve, in
the case of a large finite number of hard gas molecules. Aside from
a vanishingly small number of special initial states, the most
probable states will also occur the most frequently (at least for a
very large number of molecules). The ordinates of this curve are
almost always very small, and these small ordinates are of course
not usually maxima. It is only the ordinates with unusually large
values that are mostly maxima, and indeed they are more likely to
be maxima the greater they are. The fact that a very large ordinate
H_0 is more often a maximum than the intersection of the line
$y = H_0$ with a still higher peak is a consequence of the enormous
increase in rarity of peaks with increasing height. See the above
figure, which is of course to be taken with a pinch of salt. A correct
figure could not be printed because the H-curve actually has a large
number of maxima and minima within each finite segment, and
cannot be represented by a line with continuously changing direc-

tion. It would be better to call it an aggregate of many points very close together, or small horizontal segments.†

The Poincaré theorem is of course inapplicable to a terrestrial body which we can observe, since such a body is not completely isolated; likewise, it is inapplicable to the completely isolated gas treated by the kinetic theory, if one first lets the number of molecules become infinite, and then the quotient of the time between successive collisions and the time of observation.

Vienna, 16 December 1896.

---

† *Nature* **51,** 413, 581 (1895).

Index

247